Springer Tracts in Natural Philosophy

Volume 19

Edited by B. D. Coleman

Co-Editors:
R. Aris · L. Collatz · J. L. Ericksen · P. Germain
M. E. Gurtin · E. Sternberg · C. Truesdell

R. J. Knops · L. E. Payne

Uniqueness Theorems
in Linear Elasticity

Springer-Verlag   New York Heidelberg Berlin 1971

Robin John Knops
The University, Newcastle upon Tyne

Lawrence Edward Payne
Cornell University, Ithaca, N. Y.

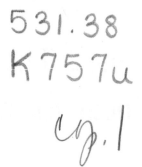

AMS Subject Classifications (1970): 35–02, 35 A 05, 73 C 15, 73–02

ISBN 0-387-05253-4 Springer-Verlag New York Heidelberg Berlin
ISBN 3-540-05253-4 Springer-Verlag Berlin Heidelberg New York

Acknowledgments

We would like to thank the Editor of this series, Professor B. D. Coleman, for inviting us to contribute a volume on uniqueness theorems in linear elasticity. We have tried to furnish a complete and up-to-date account of the subject, supported by a fairly comprehensive bibliography, but it is inevitable that omissions will occur. These omissions would have been far more serious had it not been for the kindness and willingness of a number of people to read and criticize an early version of the manuscript. For their helpful comments we wish to express our sincere gratitude to Professors B. D. Coleman, J. L. Ericksen, G. Fichera, A. E. Green and M. E. Gurtin and to Drs. M. Hayes and R. Hill. Naturally, the authors assume full responsibility for the ultimate content of this Tract.

Finally, we wish to thank the publishers for their excellence in handling the printing, and to express our appreciation to Mrs. Margaret Smith and Mrs. Jeanne Paolangeli, of Newcastle and Cornell respectively, for their help in the preparation and typing of the manuscript.

January 1971   R. J. Knops
                                                               L. E. Payne

Contents

Chapter 1

Introduction

The classical result for uniqueness in elasticity theory is due to
Kirchhoff. It states that the standard mixed boundary value problem
for a homogeneous isotropic linear elastic material in equilibrium and
occupying a bounded three-dimensional region of space possesses at
most one solution in the classical sense, provided the Lamé and shear
moduli, λ and μ respectively, obey the inequalities $(3\lambda+2\mu)>0$ and
$\mu>0$. In linear elastodynamics the analogous result, due to Neumann,
is that the initial-mixed boundary value problem possesses at most one
solution provided the elastic moduli satisfy the same set of inequalities
as in Kirchhoff's theorem. Most standard textbooks on the linear
theory of elasticity mention only these two classical criteria for uniqueness
and neglect altogether the abundant literature which has appeared since
the original publications of Kirchhoff. To remedy this deficiency it seems
appropriate to attempt a coherent description of the various contributions
made to the study of uniqueness in elasticity theory in the hope that such
an exposition will provide a convenient access to the literature while at
the same time indicating what progress has been made and what problems
still await solution. Naturally, the continuing announcement of new
results thwarts any attempt to provide a complete assessment.

Apart from linear elasticity theory itself, there are several other
areas where elastic uniqueness is significant. By way of introduction, it
is worthwhile to briefly remark on some of these inter-relations, and at
the same time to identify the position occupied by elastic uniqueness in
the more extensive field of modern continuum mechanics.

A concept usually closely associated with uniqueness is that of
stability. Current interest centers on discovering the precise nature of
the association, but the issue is complicated because of the different
concepts of stability that are employed and the fact that uniqueness
may refer to either the equilibrium or the dynamic problem. It has, in
fact, already been established that stability in the sense of Liapunov implies uniqueness of the dynamic solution, this implication holding for
general systems and not necessarily elastic ones. It might also be possible
to establish similar relations between Liapunov stability and the unique-

ness of the solution to the corresponding equilibrium problem. Such relations would obviously depend upon the type of boundary value problem under consideration, since, for example, (as is shown later) conditions for uniqueness in the displacement and traction boundary value problems differ markedly from one another. The question has not been exhaustively examined and still awaits thorough investigation.

The relation between Hadamard's definition of stability and uniqueness in the equilibrium problem has been discussed by Ericksen and Toupin [1956]. They show that stability in the Hadamard sense implies uniqueness of the displacement boundary value problem but not of the traction boundary value problem. Hill [1957] has investigated the relationship when the energy criterion is adopted as the definition of stability. It then follows that stability implies uniqueness of solution for standard boundary value problems.

Early writers on this topic, adopting the energy criterion as their stability definition, believed that the necessary and sufficient conditions for stability were provided by the classical uniqueness inequalities. This had the effect of discouraging any expansion of the original uniqueness results since it was argued that materials with moduli outside the classical range could support only unstable deformations which, whether unique or not, were of no physical interest. Of course, this view is no longer held but it probably initiated the fashion of referring only to the classical result.

Before these early ideas gained general acceptance, however, several researchers had successfully extended the original inequalities. Prominent among these were the Cosserat brothers who, endeavouring to construct a new method of solving the elastic problem, presented many new results on uniqueness in the homogeneous isotropic equilibrium problem. These results are described in the following pages and were well known by contemporary workers, being fully documented, for instance, by Appell in his book *Traité de mecanique rationelle*, which appeared in 1903. However, for the reasons suggested before, such early extensions of the classical results were gradually overlooked in the elasticity literature and then apparently entirely forgotten.

Interest in extensions of the classical theory was again revived during the post-war period of general activity in continuum mechanics, when it was realized that stability is sometimes too severe a limitation to impose upon a system. There exist large groups of important practical problems, for instance those concerned with buckling and the bursting of balloons, where it is essential to admit unstable deformations. These problems are properly studied in the context of finite elasticity, but to fully understand the destabilising mechanism, some knowledge of the uniqueness properties of the linear elastic problem is required. This has led to the current examination of the relation between the two concepts mentioned earlier.

A similar situation occurs in the development of the theory of well-posed problems in the sense defined by Hadamard. Originally, it was believed that any problem whose solution did not depend continuously upon the data—a well known example being the Cauchy problem for the Laplace equation—was of no physical significance, and hence the study of non-well-posed problems was neglected. However, it has recently been accepted that such problems, properly interpreted, do in fact provide useful models of physical situations, so that they are now the object of intensive study. The reader interested in this topic is directed to the articles by Lavrentiev [1962, 1967], John [1955], Pucci [1955] and Payne [1966].

An unsolved problem in finite elasticity is that of deciding what restrictions should be imposed upon the constitutive relations in order to produce reasonable physical behaviour. The basic thermodynamical principles supply little guidance and recourse must be made to other means. One approach has been to postulate an inequality and then show that its implications are consistent with behaviour to be expected physically. An example of this procedure is given by the inequality proposed by Coleman and Noll [1959]. A full discussion is presented by Truesdell and Noll [1965], while further assessment of this and other inequalities is given by Hill [1968, 1970]. These inequalities are, however, merely ad hoc assumptions. Accordingly, it has recently been suggested that one should examine restrictions on the constitutive relations which arise directly from the imposition of certain conditions on the deformation itself. This approach, which has been employed with respect to stability in a broad category of situations by Coleman and Mizel [1967, 1968], is the one most obviously suited to our study of uniqueness.

One point should be emphasised, however, in carrying over this approach to the study of uniqueness. It has long been recognised by all workers in the field that the requirement of a unique solution in finite elasticity is too severe. A frequently used illustrative example is that of the inversion of a hemispherical shell. Even in problems which do not admit buckling there is no clear idea of just how uniqueness should depend on the data and the geometry. It is only in the allied linear theory of small elastic deformations superposed upon large ones that our understanding of necessary and sufficient conditions approaches completion. This does not mean that no information can be derived about constitutive assumptions in the general theory. In the theory of small deformations superposed upon large, the quantities analogous to the elasticities of the classical theory are determined in part by the material properties of the finitely deformed body. Thus, for those large deformations with a small superposed deformation the restrictions imposed by uniqueness in the linear theory provide restrictions on the constitutive assumptions of the finite theory.

In spite of our assumption throughout the Tract that the various solutions under discussion exist, there is nevertheless a close relation between existence and uniqueness. This is already indicated by the Fredholm alternative. In fact we prove later that non-uniqueness of the solution to one problem—either dynamic or static—implies the non-existence of the solution of a certain dual problem. The conclusion we establish generalises slightly the original contributions of Ericksen [1963, 1964, 1965].

Described above are a few situations in which it is important to possess necessary and sufficient conditions for the uniqueness of solutions of problems in linear elasticity theory. Most of the situations are of comparatively recent origin, but it remains to note one other (probably introduced before all the rest) which may have prompted the first studies in uniqueness. Saint-Venant in his famous memoire on the torsion, bending, and flexure of beams describes how a solution he had obtained by semi-inverse methods was criticised on the grounds that it might not be complete. To meet this challenge, Saint-Venant in his memoire endeavors to show that the problem under consideration possesses a unique solution so that any expression satisfying both the data and the field equations must necessarily represent a complete solution. His proof of uniqueness is faulty but this method of justification is not. It is still a much used technique for proving completeness of solutions obtained by semi-inverse methods or by any other ad hoc precedure. One such procedure, for instance, is often used in treating problems for the half-space. Several authors have succeeded in reducing specific half-space problems to ones in potential theory. An obvious way of guaranteeing that a complete solution has been determined is by appeal to the corresponding uniqueness theorems. Remarkably, theorems for the half-space have only recently appeared and, as will be seen in the following pages, these theorems impose much weaker restrictions on the elasticities than the corresponding theorems for domains of arbitrary shape.

In this Tract we confine attention to the general problems of two- and three-dimensional linear elasticity theory, excluding from consideration such specialised topics as the torsion and flexure of beams and the deformation of rods, plates and shells. Likewise, uniqueness under point singularities is not mentioned,[1] nor is uniqueness for problems in

---

[1] Problems concerned with point singularities, while of great importance, require a different approach for uniqueness from that needed in the broad category of problems treated here. Comprehensive treatments of this topic are available, and the interested reader is referred to the articles by Sternberg and Eubanks [1955] and Sternberg and Turteltaub [1968] for the static theory, and by Sternberg and Wheeler [1968] for the dynamic theory.

thermoelasticity and dislocation theory.[2] Thus, we concentrate on non-homogeneous anisotropic linear elasticity and examine uniqueness for both weak and classical solutions to several initial and boundary value problems, giving special results for homogeneous bodies with various degrees of material symmetry, and emphasising in particular the isotropic case. Our aim is to record as completely as possible the various contributions that have been made to this subject; we do not, however, pretend to have achieved an exhaustive account.

Conditions sufficient for uniqueness are described in terms of limitations on the elasticities, regularity of solutions, and geometry of the region occupied by the elastic body, with indications of how these limitations are affected by the prescribed data. It is not possible to determine necessary conditions for uniqueness involving all of these factors. We have in fact limited our attention almost exclusively to conditions on the elasticities.

Of the various devices used to establish uniqueness theorems, the most frequently employed are those based upon energy arguments. Other methods involve reflexion principles,[3] the uniqueness theorem of Holmgren, decomposition formulae for complete representations of solutions, and logarithmic convexity arguments. Only one proof of each theorem is discussed, although where a number of variants are available we attempt to give a full list of references. Most of our proofs have appeared already in the literature, but when no acknowledgement is given, it means that a theorem or its proof appears here for the first time.

The formulations of the initial-boundary value problems to be studied are presented in Section 2.1 of Chapter 2, together with an explanation of our notation. Classical and weak solutions are defined in Section 2.2, and in Section 2.3, a summary is given of the various inequalities which we shall impose upon the elasticities in our uniqueness studies. A necessarily brief description of how these properties arise in other contexts of continuum mechanics is also included.

An essay on the early developments in elastic uniqueness constitutes the whole of Chapter 3. It contains an account of work by Kirchhoff, Neumann, the Cosserats, and Almansi, amongst others. We there indicate the extent to which later results were anticipated during this early period, especially by the Cosserats.

---

[2] Uniqueness in these problems is not considered, because the few available results are easy extensions of ones well-known in the isothermal theory. This remark does not apply to the paper by Dafermos [1968], where weak thermoelastic solutions are dealt with, nor to the work of Brun [1965, 1969].

[3] Mainly those due to Duffin which continue solutions across a plane boundary.

The exposition of modern work commences in Chapter 4. This is a substantial chapter treating the uniqueness of both weak and classical solutions to the displacement, traction, and certain mixed boundary value problems of anisotropic and isotropic, non-homogeneous and homogeneous linear elastic materials in equilibrium. Both necessary and sufficient conditions are investigated, and it is shown that in several instances relaxation of the classical criteria is possible. For example, in the displacement boundary value problem for homogeneous media the appropriate criteria reduce to the strong ellipticity condition, while in the traction boundary value problem for homogeneous isotropic media a different criterion relaxation is possible. Consideration is also given to uniqueness in exterior domains. Here, the main difficulty lies in specifying for the stress or displacement a reasonable behaviour at infinity which will guarantee the asymptotic decay essential for the convergence of pertinent integrals. In separate sections we record interesting special results for homogeneous isotropic bodies bounded internally by a spherical surface.

Mixed boundary value problems with four different types of boundary conditions are considered in Section 4.4. We first obtain results for a general non-homogeneous anisotropic body, and then consider in detail some special problems for homogeneous isotropic bodies occupying regions of various geometrical shapes. In particular, uniqueness criteria are given for bodies bounded by two parallel planes—the infinite slab—and also by a plane and a conical surface.

The next chapter describes uniqueness theorems for plane strain equilibrium problems and concentrates attention on the homogeneous isotropic body. It is shown that, for the traction problem, necessary and sufficient conditions for uniqueness are $\sigma \neq 1$, $\mu \neq 0$ (where σ is Poisson's ratio), but that a similar improvement on the classical result cannot be made in the displacement boundary problem except for bodies with special geometries. The proofs of this chapter do not make use of complex analysis. The chapter closes with an appendix on related results for axisymmetric solutions.

Chapter 6 is devoted to problems for the whole or half space and contains results in the corresponding displacement, traction and mixed boundary value problems. We first establish essential estimates on the asymptotic behaviour of the stress and displacement in a neighborhood of infinity, and then make use of reflexion principles to obtain uniqueness theorems in all cases except that of the "smooth-rigid-die" mixed problem for which we employ a decomposition formula. In each case conditions on the elasticities, both necessary and sufficient for uniqueness, are given. For the displacement problem these are $\sigma \neq \frac{3}{4}$, 1, $\mu \neq 0$, while in the remaining problems they are $\sigma \neq 1$, $\mu \neq 0$.

There next follows in Chapter 7 a brief account of some miscellaneous problems associated with a sphere. In this chapter we also give a new proof of a theorem, originally due to Almansi, concerned with the prescription of Cauchy data—the displacement and traction—on a portion of the surface. A final section discusses certain "ambiguous" boundary conditions introduced by Signorini [1959a, b] and considered in detail by Fichera [1963a, b]. Although the problems of this chapter are unconventional, they nevertheless are not without practical interest, e. g. in punch problems and in certain problems in geology and geophysics.

The final chapter deals with dynamic problems. It is possible in this case to dispense with the variety of conditions which we required for uniqueness in the static problems, since uniqueness for all the standard initial-boundary value problems is guaranteed by the simple requirement that the governing differential equation be self-adjoint, i.e. that the elasticities possess major symmetry. The result is valid for non-homogeneous anisotropic bodies in bounded regions. Exterior regions are also considered, but only for homogeneous bodies. In the concluding sections of this chapter we examine some of the relations among the concepts of uniqueness, boundedness, stability and existence. This examination is confined principally to the dynamic problem although some relations hold also for the equilibrium solution.

We end the introduction with an indication of some problems thus far unsolved. These may be conveniently classified into four groups:

I. In the classical theory of isotropic bodies one observes that if $\mu \neq 0$, then for the usual types of boundary value problems there is a discrete set of values of Poisson's ratio for which the equilibrium problem has a non-unique solution. These values vary with geometry and with the type of boundary value problem considered. Thus in the case of anisotropic elasticity one would expect that for special geometries the necessary and sufficient conditions for uniqueness might be quite different from those required for the general case. We shall observe this already in the case of the isotropic sphere, but it would be of interest to examine other bodies occupying regions with such simple shapes as the cube, quarterplane and half-plane (or slab) pierced by a cylindrical hole.

II. In the equilibrium traction boundary value problem for homogeneous isotropic media occupying bounded domains, a counter-example to uniqueness for values of Poisson's ratio greater than unity has not been found. The known counter-examples concern regions of unbounded extent. The absence of such counter-examples strengthens the belief that for interior problems uniqueness holds provided $\sigma > -1$, $\sigma \neq 1$, and for exterior problems uniqueness holds provided $\sigma < 1$. This conjecture arises quite naturally from certain theorems and examples discussed fully in the text.

Chapter 2

Basic Equations

The basic equations studied in the Tract are recorded in this chapter together with a summary of the different types of initial-boundary value problems that are considered. We briefly indicate possible situations in which these problems arise and establish some general properties of their solutions. Classical and generalised solutions are defined here, and a short description is given of the various types of inequalities which are to be placed on the elasticities.

2.1 Formulation of Initial-Boundary Value Problems

We suppose that a deformed elastic body occupies an open region B of euclidean three-(or two-)space with a bounding surface ∂B that is piecewise smooth. We denote the closure of B by $\bar{B}$. Usually the surface ∂B is taken to be closed and bounded with the region B lying either internal or external to it, but exceptionally ∂B may be taken as open and extending to infinity. We also consider the situation in which ∂B is located entirely at infinity, the body then occupying the whole space.

In the following formulation of initial-boundary value problems, we discuss first the three-dimensional theory and afterwards introduce those modifications needed for the two-dimensional theory.

Throughout the Tract we adopt the convention of summing over repeated suffixes, and we always use a fixed cartesian coordinate system.

The deformations of the body are measured from some reference configuration, and when referred to the fixed axes, the components of the Cauchy stress tensor σ_{ij} in the deformed configuration of B are known to satisfy the relations

$$\left.\begin{aligned}\frac{\partial \sigma_{ij}}{\partial x_j} + \rho F_i = \rho \frac{\partial^2 u_i}{\partial t^2}\\ \sigma_{ij} = \sigma_{ji}\end{aligned}\right\} \quad \text{in} \quad B \times (0, T), \quad i, j = 1, 2, 3, \qquad \begin{aligned}(2.1.1)\\[1em](2.1.2)\end{aligned}$$

in which $u_i(\mathbf{x}, t)$ are the cartesian components of the displacement, $F_i(\mathbf{x}, t)$ the components of the body force per unit mass, T some prescribed value of time (for most purposes chosen to be $+\infty$) and $\rho(\mathbf{x})$ the density, taken to be positive. Since our principal concern is with properties of the linear equations of infinitesimal elasticity, we may identify the coordinates x_i with the positions of material particles in either the reference or deformed configuration of the body. Thus, to the order of approximation envisaged by the infinitesimal theories, the stress σ_{ij} in (2.1.1) may be referred to coordinate surfaces in either configuration.

In the classical linear theories of elasticity, the natural configuration of the body, i.e. that in which both the stress and strain vanish, is taken as the reference configuration. The stress is related linearly to the linear strain, whose cartesian components are given by

$$e_{ij} = \frac{1}{2}\left(\frac{\partial u_i}{\partial x_j} + \frac{\partial u_j}{\partial x_i}\right),\tag{2.1.3}$$

in one of two possible ways. Either it is directly postulated that the stress is a linear function of these strains, or the existence of a homogeneous quadratic strain energy function is assumed. In the first approach, which originated with Cauchy, one lays down the relation

$$\sigma_{ij} = c_{ijkl}\, e_{kl},\tag{2.1.4}$$

which, because of the symmetry of both stress and strain, immediately imposes upon the *elasticities*, $c_{ijkl}(\mathbf{x})$, the symmetry requirements

$$c_{ijkl} = c_{jikl} = c_{ijlk}.\tag{2.1.5}$$

The elasticities are material functions of position which, in particular, may be constant. If they are constant then we say that the material is *homogeneous*. For a body without material symmetry, there are thirty-six independent elasticities. This number is progressively reduced as the material symmetry of the body increases, until for an isotropic linear elastic material only two remain. Then we have

$$c_{ijkl} = \lambda\,\delta_{ij}\,\delta_{kl} + \mu(\delta_{ik}\,\delta_{jl} + \delta_{il}\,\delta_{jk}),\tag{2.1.6}$$

which expresses the fact that the elasticities are now the components of an isotropic tensor. In (2.1.6), δ_{ij} represents the Kronecker delta, which equals 1 if $i=j$, and is zero otherwise. The elasticities $\lambda(\mathbf{x})$ and $\mu(\mathbf{x})$ entering into (2.1.6) are known respectively as the *Lamé* and *shear moduli*. They are related to *Poisson's ratio* σ and *Young's modulus* E by

$$\lambda = 2\mu\sigma/(1-2\sigma); \quad \mu = E/2(1+\sigma).\tag{2.1.7}$$

Insertion of (2.1.6) into (2.1.4) leads to Hooke's law:

$$\sigma_{ij} = \lambda\, e_{kk}\,\delta_{ij} + 2\mu\, e_{ij}.\tag{2.1.8}$$

The second definition of classical linear elasticity rests upon the postulate, proposed by Green, that *the work done by the stress in a deformation depends only upon the strain and is recoverable work.* (See Truesdell and Toupin [1960, p. 724].) After appropriate linearisation, this concept leads to a stress-strain relation identical in form to (2.1.4) but with the elasticities possessing the additional symmetry

$$c_{ijkl} = c_{klij}. \tag{2.1.9}$$

The number of independent elasticities is therefore reduced to twenty-one. In the case of isotropic bodies there is clearly no distinction between the two definitions, but for anisotropic bodies a distinction exists and becomes crucial in questions of uniqueness.

While we may consider these theories of infinitesimal elasticity in their own right, we may also regard them, with suitable modifications, as approximations to certain non-linear theories. The reader interested in such linearisation procedures will find a full account in the monograph by Truesdell and Noll [1965],[1] but here we content ourselves with mentioning only the theory of small elastic deformations superposed upon large arbitrary static elastic deformations. The reference configuration is that of the body in its state of initial large deformation and $\mathbf{x}$ now denotes the position vector in this state. By postulating the existence of a strain energy function, it may be proved that in the small additional deformation the stress $\sigma_{ij}(\mathbf{x}, t)$ associated with the small incremental displacement $u_i(\mathbf{x}, t)$ is given by

$$\sigma^{ij} = \bar{c}_{ijkl} \frac{\partial u_k}{\partial x_l} \quad \text{in} \quad B \times (0, T), \tag{2.1.10}$$

$$\bar{c}_{ijkl} = \bar{c}_{klij}. \tag{2.1.11}$$

In (2.1.10), σ_{ij} denotes the first Piola-Kirchhoff (or nominal) stress based on the reference configuration. This stress continues to satisfy the equations of motion (2.1.1), but is no longer symmetric.

The quantities $\bar{c}_{ijkl}(\mathbf{x})$ entering into (2.1.10) may be identified with the elasticities introduced above (hence the choice, apart from the superposed bar, of identical symbols). However, the $\bar{c}_{ijkl}$ bear an intrinsic dissimilarity to the quantities c_{ijkl} because they depend not only upon the material but also upon the initial large strain. Only when it is assumed that the body is initially homogeneous and also that the finite static strain is homogeneous are the $\bar{c}_{ijkl}$ constant. Henceforth, we omit the bar when

---

[1] For the reduction of simple materials with memory, see p. 117: for elastic fluids, see p. 151; for elastic bodies, referred to in the text above, see p. 246, and Green, Rivlin and Shield [1952].

discussing equation (2.1.10), and broaden the term "elasticities" to include the moduli of (2.1.10).

It is also worth remarking that when invariance principles are applied to the theory, the $\bar{c}_{ijkl}$ are required to satisfy extra conditions which are noted below in (2.1.17). However, apart from (2.1.11), no other symmetries need be imposed on these quantities.

Insertion of the respective constitutive expressions (2.1.4) and (2.1.10) into the equations of motion (2.1.1) leads to the equations governing the displacement:

$$\frac{\partial}{\partial x_j}\left(c_{ijkl}\frac{\partial u_k}{\partial x_l}\right)+\rho F_i=\rho\frac{\partial^2 u_i}{\partial t^2}\quad\text{in }B\times(0,T),\qquad(2.1.12)$$

where we have presupposed the validity of the indicated differentiations. In the theory of small deformations superposed upon large deformations, the position vector $\mathbf{x}$ and the density refer to the reference configuration, while the vector $\mathbf{F}$ is given by

$$F_i=f_i+\frac{1}{\rho}\frac{\partial\sigma_{ij}^0}{\partial x_j}\qquad(2.1.13)$$

where σ_{ij}^0 is the Cauchy stress in the initial large deformation, and f_i is the small increment in body force.

The various cases considered so far are distinguished in (2.1.12) only by the respective symmetries required of the elasticities. These symmetries are now listed for convenience.

I. Classical "Cauchy" elasticity:

$$c_{ijkl}=c_{jikl}=c_{ijlk}.\qquad(2.1.14)$$

II. Classical "Green" elasticity:

$$c_{ijkl}=c_{jikl}=c_{klij}.\qquad(2.1.15)$$

III. Small deformations superposed upon large (Green):

$$c_{ijkl}=c_{klij}.\qquad(2.1.16)$$

In III, when the final Cauchy stress is made properly invariant, the elasticities then satisfy the further requirements:

$$c_{ijkl}-\delta_{ik}\sigma_{jl}^0=c_{jikl}-\delta_{jk}\sigma_{il}^0=c_{ijlk}-\delta_{il}\sigma_{jk}^0\qquad(2.1.17)$$

where, as above, σ_{ij}^0 is the Cauchy stress of the initial large deformation. Setting $\sigma_{ij}^0=0$ shows immediately that the elasticities satisfy the symmetries (2.1.15) of the classical Green theory.

There is one other theory to be considered, namely the less restrictive concept of elasticity which does not involve a strain energy function. The equations governing the small incremental displacement and appropriate to a theory slightly more general than elasticity (in the sense that it is not properly invariant) are again given by (2.1.12) and (2.1.13), but now *there are no symmetry requirements* on the c_{ijkl}. Introduction of the pertinent invariance requirements shows, however, that condition (2.1.17) must be satisfied. In particular if $\sigma_{ij}^0 = 0$, this condition reduces the c_{ijkl} to constants obeying the symmetries (2.1.14) of classical Cauchy elasticity.

To complete the definition of the initial-boundary value problem we adjoin to (2.1.12) the initial conditions (the symbol $B(t)$ refers to the body B at time t)

$$u_i(\mathbf{x}, 0) = f_i(\mathbf{x}), \qquad \frac{\partial u_i}{\partial t}(\mathbf{x}, 0) = g_i(\mathbf{x}) \qquad \text{on } B(0), \qquad (2.1.18)$$

and the boundary conditions

$$u_i(\mathbf{x}, t) = h_i(\mathbf{x}, t) \qquad \text{on } \overline{\partial B_1} \times [0, T), \qquad (2.1.19)$$

$$n_j \sigma_{ij} \equiv n_j c_{ijkl} \frac{\partial u_k}{\partial x_l}(\mathbf{x}, t) = l_i(\mathbf{x}, t) \qquad \text{on } \partial B_2 \times [0, T), \qquad (2.1.20)$$

where f_i, g_i, h_i and l_i are prescribed functions, ∂B_1 and ∂B_2 are disjoint subsets of ∂B such that $\overline{\partial B_1} \cup \partial B_2 = \partial B$, and n_i is the unit outward normal on ∂B_2. Boundary conditions other than (2.1.19) and (2.1.20) are considered later, and these are explained when introduced. It is possible that either ∂B_1 or ∂B_2 may be empty. For $\partial B_1 = \emptyset$, imposition of the normalisations

$$\int_{B(t)} \rho u_i \, dx = e_{ijk} \int_{B(t)} \rho x_j u_k \, dx = 0, \qquad (2.1.21)$$

$$\int_{B(t)} \rho \frac{\partial u_i}{\partial t} \, dx = e_{ijk} \int_{B(t)} \rho x_j \frac{\partial u_k}{\partial t} \, dx = 0, \qquad (2.1.22)$$

excludes arbitrary rigid body motions. Here, e_{ijk} is the permutation symbol, defined to be $+1$ or -1 if i, j, k is an even or odd permutation of 1, 2, 3, respectively, and zero otherwise. When the elasticities possess only the major symmetry (2.1.16), the second condition in each set may be omitted since the rotational part of the motion is completely determined by the problem.

The conditions for an equilibrium problem are obtained by merely dropping the acceleration term in (2.1.12) and ignoring both the initial conditions (2.1.18) and the normalisation conditions (2.1.22). Thus, the governing equations are:

$$\frac{\partial}{\partial x_j}\left(c_{ijkl} \frac{\partial u_k}{\partial x_l}\right) + \rho F_i = 0 \qquad \text{in } B, \qquad (2.1.23)$$

while on ∂B, for example,

$$u_i(\mathbf{x}) = H_i(\mathbf{x}) \quad \text{on } \overline{\partial B_1},$$

$$n_j \sigma_{ij} = n_j c_{ijkl} \frac{\partial u_k}{\partial x_l}(\mathbf{x}) = G_i(\mathbf{x}) \quad \text{on } \partial B_2, \tag{2.1.24}$$

where H_i, G_i are prescribed functions, and $\partial B = \overline{\partial B_1} \cup \partial B_2$. Boundary conditions other than (2.1.24) of both standard and non-standard type are considered later in the Tract. The elasticities satisfy the symmetries appropriate to the manner in which Eqs. (2.1.23) have been derived, in particular possessing no symmetries whatsoever for the theory of small elastic deformations superposed upon large in the absence of a strain energy function. In all situations in which the elasticities satisfy the major symmetry (2.1.16), the governing system (2.1.23) is formally self-adjoint. When ∂B_1 is empty, for equilibrium the loading is constrained by

$$\int_B \rho F_i \, dx + \int_{\partial B} G_i \, dS = 0. \tag{2.1.25}$$

Since from the viewpoint of uniqueness, we deal exclusively with homogeneous data, condition (2.1.25) is automatically satisfied in our proofs.

Where the region occupied by the body is of infinite extent, the prescription of boundary conditions at infinity requires special care and is usually governed by the requirement that certain integrals vanish over the (partial) surface of spheres of infinite radius. However, in the case, for instance, of a homogeneous isotropic body in equilibrium, Fichera [1950] and Gurtin and Sternberg [1961a] (see also Chapter IV, §4.2) have shown that the *obvious* implication of such a requirement gives an artificial asymptotic decay of the displacement and stress which may, in fact, be replaced by a weaker and more realistic behaviour. We discuss this aspect further in the appropriate sections.

2.2 The Classical and Weak Solutions

The uniqueness theorems of this Tract involve classical and weak solutions of the initial-boundary value problems specified at the end of the previous section, it being assumed that a solution always exists in the appropriate sense. We now wish to define these concepts, doing so first for the dynamic problem and then for the equilibrium one.

By a classical solution we mean a vector valued function $u_i(\mathbf{x}, t) \in C^2$ in $B \times (0, T)$, satisfying (2.1.12) (subject to the pertinent symmetries on the elasticities) and the initial and boundary conditions (2.1.18), (2.1.19), and

(2.1.20). The definition implies that the indicated differentiations are justified.

The notion we employ for a weak solution is the following. Let $\mathbf{W}$ denote the set of all vector functions $\mathbf{w}(\mathbf{x}, t)$ that are continuous in $\bar{B} \times [0, T]$ and have square integrable derivatives in $B \times [0, T]$. In addition, for almost all $\mathbf{x}$, the derivatives $\partial w_i/\partial t$ and $\partial w_i/\partial x_j$ are assumed to exist as continuous functions of t for $0 < t \leq T$. Let Φ denote the set of vector functions $\boldsymbol{\varphi}(\mathbf{x}, t)$ that are continuously differentiable in $B \times [0, T]$ and that vanish on the sets $B(0)$ and $\overline{\partial B_1} \times [0, T)$. The reason for introducing a weak solution is in order to treat problems for which the data and the geometry do not possess the continuity and differentiability necessary for the existence of a classical solution. However, the data cannot be completely arbitrary, even for a weak solution, and we therefore impose the following restrictions

(a) $\rho(\mathbf{x})$ and $c_{ijkl}(x)$ are bounded measurable functions in B,

(b) $f_i(\mathbf{x})$ and $h_i(\mathbf{x}, t)$ are continuous functions on their respective regions of definitions,

(c)

$$\int_{\partial B_2 \times [0, T]} l_i l_i \, dS + \int_0^T \int_{B(\eta)} F_i F_i \, dx \, d\eta + \int_B \frac{\partial f_i}{\partial x_j} \frac{\partial f_i}{\partial x_j} \, dx + \int_B g_i g_i \, dx < \infty. \quad (2.2.1)$$

With these conditions on the data, we define our weak solution to be a vector-valued function $\mathbf{u}(\mathbf{x}, t) \in \mathbf{W}$ which for each $\boldsymbol{\varphi} \in \Phi$ and for each arbitrary $t \in [0, T)$ satisfies

$$\int_{B(t)} \rho \, \varphi_i \frac{\partial u_i}{\partial t} \, dx - \int_0^t \int_{B(\eta)} \left[\rho \frac{\partial \varphi_i}{\partial \eta} \frac{\partial u_i}{\partial \eta} - c_{ijkl} \frac{\partial \varphi_i}{\partial x_j} \frac{\partial u_k}{\partial x_l} \right] dx \, d\eta$$

$$= \int_{\partial B_2(t)} l_i \varphi_i \, dS + \int_0^t \int_{B(\eta)} \rho F_i \varphi_i \, dx \, d\eta, \quad (2.2.2)$$

$$u_i(\mathbf{x}, t) = h_i(\mathbf{x}, t) \quad \text{on } \partial B_1 \times [0, T).$$

The weak solution also must satisfy the initial conditions

$$u_i(\mathbf{x}, 0) = f_i(\mathbf{x}),$$

$$\frac{\partial u_i}{\partial t}(\mathbf{x}, 0) = g_i(\mathbf{x}), \qquad \frac{\partial u_i}{\partial x_j}(\mathbf{x}, 0) = \frac{\partial f_i}{\partial x_j}(\mathbf{x}), \qquad \text{for almost all } \mathbf{x} \in B, \quad (2.2.3)$$

and the following conservation law

$$E(t) \equiv \frac{1}{2} \int_{B(t)} \left[\rho \frac{\partial w_i}{\partial t} \frac{\partial w_i}{\partial t} + c_{ijkl} \frac{\partial w_i}{\partial x_j} \frac{\partial w_k}{\partial x_l} \right] dx = E(0), \qquad 0 \leq t \leq T, \quad (2.2.4)$$

where $E(0)$ is constant and w_i is the difference of any two solutions of (2.2.2).

It is desirable that every classical solution be also a weak solution. For the *dynamic* equations, this is so if and only if the elasticities possess the major symmetry (2.1.16), since otherwise the conservation law (2.2.4), which is both necessary and sufficient for the symmetry (2.1.16), (Truesdell [1963]) would hold for weak but not for classical solutions. No such inconsistency arises in the *static* theory. Here, the weak solution may be defined with the same restrictions (a), (b) and (c) on the appropriate data and with the same spatial continuity and differentiability properties imposed on the functions $\mathbf{w}(\mathbf{x})$, $\varphi(\mathbf{x})$ which are now dependent upon $\mathbf{x}$ alone. In the static case we define as a weak solution, a function $\mathbf{u}(\mathbf{x}) \in \mathbf{W}$, which for every $\varphi \in \boldsymbol{\Phi}$ satisfies

$$\int_B c_{ijkl} \frac{\partial \varphi_i}{\partial x_j} \frac{\partial u_k}{\partial x_l} \, dx = \int_{\partial B_2} l_i \varphi_i \, dS + \int_B \rho F_i \varphi_i \, dx. \qquad (2.2.5)$$

Then, clearly, every classical solution is a weak solution, since (2.2.5) follows from the divergence theorem and the equilibrium equation corresponding to (2.1.12), without any restriction on the symmetry of the elasticities.

2.3 The Homogeneous Isotropic Body. Plane Elasticity

Considerable simplication of the governing equations occurs for homogeneous isotropic linear elastic bodies. These equations have attracted intensive study, and it turns out that many of the uniqueness theorems in the static theory are concerned with the isotropic case. It is convenient therefore to collect in this section those parts of the theory relevant to our purposes, but restricting ourselves to classical solutions. At the end of the section we recall the plane theory.

Substitution of (2.1.6) into (2.1.12) produces the Navier equations for the displacement:

$$\mu \, \Delta u_i + (\lambda + \mu) \frac{\partial^2 u_j}{\partial x_i \partial x_j} + \rho F_i = \rho \frac{\partial^2 u_i}{\partial t^2} \qquad \text{in } B \times (0, T), \qquad (2.3.1)$$

in which $\Delta \left(= \dfrac{\partial^2}{\partial x_i \partial x_i} \right)$ is the Laplace operator. The data for the problem are merely the initial and boundary conditions of the anisotropic case. In equilibrium and in the absence of body force Eq. (2.3.1) becomes

$$\mu \, \Delta u_i + (\lambda + \mu) \frac{\partial^2 u_j}{\partial x_i \partial x_j} = 0 \qquad \text{in } B, \qquad (2.3.2)$$

the boundary conditions being those of (2.5.19) and (2.1.20). Operating on (2.3.2) with the divergence and curl yields, respectively,[2]

$$(\lambda+2\mu)\,\Delta\,\frac{\partial u_j}{\partial x_j}=0, \quad \mu\,e_{ijk}\,\Delta\,\frac{\partial u_k}{\partial x_j}=0 \quad \text{in } B, \tag{2.3.3}$$

implying that the conditions

$$(\lambda+2\mu)\neq0 \quad (\text{or } \mu\neq0,\ \sigma\neq1), \tag{2.3.4}$$

and

$$\mu\neq0, \tag{2.3.5}$$

are, respectively, sufficient to ensure the harmonicity in B of the dilatation $\frac{\partial u_i}{\partial x_i}$ and rotation $\frac{1}{2}e_{ijk}\frac{\partial u_k}{\partial x_j}$. An application of the Laplace operator to (2.3.2) then shows that (2.3.4) and (2.3.5) guarantee the biharmonicity in B of the displacement, and therefore of the strain and stress. Particular cases of this theory corresponding to special choices of λ and μ (e.g. incompressibility) are considered at appropriate points in the text.[3]

In the theory of *infinitesimal plane strain*, the displacement is

$$u_\alpha=u_\alpha(x_1,x_2,t), \quad \alpha=1,2$$
$$u_3\equiv0, \tag{2.3.6}$$

the constitutive relations (2.1.4) therefore becoming

$$\sigma_{ij}(x_1,x_2,t)=c_{ij\alpha\beta}\frac{\partial u_\alpha}{\partial x_\beta}, \quad \alpha,\beta=1,2. \tag{2.3.7}$$

Under the assumptions of homogeneity and isotropy, (2.3.7) reduces to

$$\sigma_{\alpha\beta}=\lambda\,e_{\gamma\gamma}\delta_{\alpha\beta}+2\mu\,e_{\alpha\beta}, \quad \gamma=1,2,$$
$$\sigma_{33}=\lambda\,e_{\gamma\gamma}$$
$$\sigma_{13}=\sigma_{23}=0, \tag{2.3.8}$$

where $e_{\alpha\beta}=\frac{1}{2}\left(\frac{\partial u_\alpha}{\partial x_\beta}+\frac{\partial u_\beta}{\partial x_\alpha}\right)$; the equations of motion become

$$\frac{\partial\sigma_{\alpha\beta}}{\partial x_\beta}+\rho F_\alpha=\rho\frac{\partial^2 u_\alpha}{\partial t^2}, \tag{2.3.9}$$
$$F_3\equiv0.$$

---

[2] Duffin's [1956] result justifies the differentiations for $\lambda+2\mu\neq0$. See also Fichera [1965].

[3] For incompressibility ($\sigma=\frac{1}{2}$), see p. 48; for $\sigma=1$, $\mu\neq0$, see p. 28; for $\sigma=-1$, $\mu\neq0$, see p. 41; for $\sigma\neq1$, -1, $\mu=0$, see p. 29.

The corresponding Navier equations are

$$\mu \Delta u_\alpha + (\lambda + \mu) \frac{\partial^2 u_\beta}{\partial x_\alpha \partial x_\beta} + \rho F_\alpha = \rho \frac{\partial^2 u_\alpha}{\partial t^2}, \qquad (2.3.10)$$

which for the static theory in the absence of body force give

$$\mu \Delta u_\alpha + (\lambda + \mu) \frac{\partial^2 u_\beta}{\partial x_\alpha \partial x_\beta} = 0. \qquad (2.3.11)$$

Upon rearranging (2.3.11) we therefore get

$$(\lambda + 2\mu) \frac{\partial e_{\gamma\gamma}}{\partial x_1} = 2\mu \frac{\partial \omega}{\partial x_2}, \qquad (\lambda + 2\mu) \frac{\partial e_{\gamma\gamma}}{\partial x_2} = -2\mu \frac{\partial \omega}{\partial x_1}, \qquad (2.3.12)$$

where the rotation ω is given by $\frac{1}{2} e_{3\alpha\beta} \frac{\partial u_\beta}{\partial x_\alpha}$ and Greek subscripts take the values 1, 2. We recognise immediately that Eqs. (2.3.12) are the Cauchy-Riemann relations for the dilatation and rotation. They are therefore harmonic in B, provided we are considering the classical solution and restrict the moduli to satisfy $\sigma \neq 1$, $\mu \neq 0$. From (2.3.11) we may now infer that the displacement is biharmonic under the same assumptions.

2.4 Definiteness Properties of the Elasticities

Many of the theorems proved in the Tract depend upon the specific definiteness conditions imposed on the elasticities. Listed in this section are all the required definitions together with a few remarks concerning their relations with other topics. Note that most definitions require no symmetry or homogeneity of the elasticities.

I. *Positive Definite.* The elasticities are positive-definite when the inequality

$$c_{ijkl} \xi_{ij} \xi_{kl} > 0 \qquad (2.4.1)$$

is satisfied at all points in B for arbitrary non-zero tensors ξ_{ij}. Alternatively, we may define positive-definiteness of the elasticities to mean that at all points of B the inequality[4]

$$c_{ijkl} \xi_{ij} \xi_{kl} \geq c \, \xi_{ij} \xi_{ij} \qquad (2.4.2)$$

is satisfied for all non-zero tensors ξ_{ij} and for some positive constant c. We note that (2.4.1) and (2.4.2) are equivalent when the elasticities and the tensors ξ_{ij} are continuous functions with compact domain. For

---

[4] Compare e.g., Fichera [1965, p. 102].

elasticities obeying the symmetries

$$c_{ijkl} = c_{jikl} = c_{ijlk} \tag{2.4.3}$$

inequalities (2.4.1), (2.4.2) clearly cannot hold in the unrestricted class of tensors ξ_{ij} [5] but only in the symmetric class $\xi_{ij} = \xi_{ji} \not\equiv 0$. We then have (2.4.2) replaced by

$$c_{ijkl} \eta_{ij} \eta_{kl} \geqq c_1 \eta_{ij} \eta_{ij}, \qquad \eta_{ij} = \eta_{ji}, \tag{2.4.4}$$

for some positive constant c_1.

The isotropic elastic tensor (2.1.6) satisfies the symmetry (2.4.3) and hence is positive-definite in the class of symmetric tensors provided (2.4.4) is fulfilled. In three dimensions, necessary and sufficient conditions are

$$(3\lambda + 2\mu) > 0, \qquad \mu > 0$$

or equivalently

$$-1 < \sigma < \tfrac{1}{2}, \qquad \mu > 0. \tag{2.4.5}$$

On the other hand, in two dimensions necessary and sufficient conditions are

$$(\lambda + 2\mu) > 0, \qquad \mu > 0,$$

or equivalently

$$-\infty < \sigma < \tfrac{1}{2}, \qquad 1 < \sigma < \infty, \qquad \mu > 0. \tag{2.4.6}$$

II. *Positive Semi-Definite*. The elasticities are positive semi-definite when the inequality

$$c_{ijkl} \xi_{ij} \xi_{kl} \geqq 0 \tag{2.4.7}$$

is satisfied at all points of B for arbitrary non-zero tensors ξ_{ij}. In the three-dimensional isotropic case necessary and sufficient conditions for (2.4.7) to hold are

$$(3\lambda + 2\mu) \geqq 0, \qquad \mu \geqq 0,$$

or equivalently

$$-1 \leqq \sigma < \tfrac{1}{2}, \qquad \mu \geqq 0, \tag{2.4.8}$$

while in the two-dimensional isotropic case the corresponding necessary and sufficient conditions are

$$(\lambda + 2\mu) \geqq 0, \qquad \mu \geqq 0,$$

or equivalently

$$-\infty < \sigma < \tfrac{1}{2}, \qquad 1 \leqq \sigma < \infty, \qquad \mu \geqq 0. \tag{2.4.9}$$

It is immaterial to the definition (2.4.7) what symmetries (if any) are imposed on the elasticities.

---

[5] Inequalities (2.4.1), (2.4.2) fail, for instance whenever $\xi_{ij} = -\xi_{ji}$.

III. *Negative Definite and Negative Semi-Definite.* These properties are defined analogously to I and II but with the inequalities reversed.

IV. *Ellipticity.* The equilibrium equations (2.1.23), i.e.,

$$\frac{\partial}{\partial x_j}\left(c_{ijkl}\frac{\partial u_k}{\partial x_l}\right)+\rho F_i=0, \tag{2.1.23}$$

are elliptic in B if the elasticities satisfy at all points of B the condition

$$\det(c_{ijkl}\xi_j\xi_l)\neq 0, \tag{2.4.10}$$

for every non-vanishing real vector ξ_i. In the case of isotropy, (2.4.10) reduces to (cf. e.g. Zorski [1962]),

$$(\lambda+2\mu)\neq 0, \quad \mu\neq 0,$$

or
$$\sigma\neq 1, \quad \mu\neq 0. \tag{2.4.11}$$

The same condition holds in the two-dimensional isotropic problem.

V. *Strong Ellipticity.* The equilibrium Eqs. (2.1.23) are strongly elliptic in B if at every point of B the elasticities satisfy the condition

$$c_{ijkl}\eta_i\eta_k\xi_j\xi_l\neq 0 \tag{2.4.12}$$

for all non-zero vectors ξ_i and η_i. In connected regions a necessary and sufficient condition for (2.4.12) to hold is that the biquadratic form on the left be definite—positive or negative—i.e., either

$$c_{ijkl}\eta_i\eta_k\xi_j\xi_l>0, \quad \xi_i,\eta_i\neq 0, \tag{2.4.13}$$

or
$$c_{ijkl}\eta_i\eta_k\xi_j\xi_l<0, \quad \xi_i,\eta_i\neq 0, \tag{2.4.14}$$

imply the strong ellipticity of (2.1.23).

We may alternatively replace (2.4.13) by the inequality

$$c_{ijkl}\eta_i\eta_k\xi_j\xi_l\geq c_2\eta_i\eta_i\xi_j\xi_j \tag{2.4.15}$$

for some positive constant c_2, with an analogous inequality replacing (2.4.14). The alternate postulates are equivalent to (2.4.13) and (2.4.14) when the elasticities and vector functions η_i and ξ_i are continuous functions with compact domain. From either (2.4.13) or (2.4.15) we see that strong ellipticity is a weaker condition to impose than positive definiteness because the tensors composed from $\eta_i\xi_j$ are a subclass of second order tensors. Similarly, strong ellipticity is weaker than negative definiteness. Observe further that under the assumption of the symmetries (2.4.3), the positive-definiteness inequality (2.4.4) implies the strong ellipticity inequality (2.4.15). By virtue of (2.4.13) or (2.4.14) strong ellipticity may also be interpreted as showing that the quadratic form associated with $c_{ijkl}\eta_i\eta_k$ is definite, hence implying that *strong ellipticity implies ellipticity*. The converse is not true.

For isotropy the condition of strong ellipticity leads to the result (cf., e.g., Zorski [1962]),

$$\mu(\lambda+2\mu)>0,$$

i.e.,   (2.4.16)

$$-\infty<\sigma<\tfrac{1}{2}, \quad 1<\sigma<\infty, \quad \mu\neq0,$$

which is valid for either three or two dimensions.

VI. *Semi-Strong Ellipticity.* The equilibrium Eqs. (2.1.23) are semi-strongly elliptic in B provided the elasticities satisfy at all points of B the condition

$$c_{ijkl}\eta_i\eta_k\xi_j\xi_l\geq0, \qquad (2.4.17)$$

for all non-zero vectors ξ_i and η_i. This condition, obviously weaker than positive semi-definiteness, reduces in the isotropic case to

$$\mu(\lambda+2\mu)\geq0,$$

i.e.,   (2.4.18)

$$-\infty<\sigma\leq\tfrac{1}{2}, \quad 1\leq\sigma<\infty, \quad \mu\neq0$$

which holds in either two or three dimensions.

As mentioned above, several of the following uniqueness theorems employ the definitions we have just given. However, a close inspection of the proofs reveals that some of these theorems actually require slightly weaker assumptions corresponding to the integral forms of the respective inequalities. For instance, in the proof of the classical Kirchhoff theorem it is sufficient to assume the inequality

$$\int_B c_{ijkl}\xi_{ij}\xi_{kl}\,dx\geq c_3\int_B \xi_{ij}\xi_{ij}\,dx, \qquad (2.4.19)$$

valid for symmetric second order tensors ξ_{ij} and some positive constant c_3, instead of the more stringent condition (2.4.4). It is obvious from the context where such relaxations can be allowed, but for simplicity all theorems employ the definitions I–VI.

Although none of the restrictions listed under I–VI can be justified from a thermodynamical standpoint, several attempts have been made to show that they lead to reasonable consequences. In the classical infinitesimal isotropic theory, Truesdell and Toupin [1965][6] related the positive definite criteria (2.4.5) to a number of physically reasonable properties that it was felt an elastic body should possess. The analysis was extended to finite deformations, and consequences of strong ellipticity

---

[6] See also Truesdell and Noll [1965 §51].

were examined and compared with those of an inequality proposed by Coleman and Noll [1959].[7] Hill [1962] gave an interpretation of the strong ellipticity conditions equally valid in terms of steady velocity fields or in the context of elastostatics, whereas Zorski [1962], Shield [1965], and Knops and Wilkes [1966] explored the significance of the condition for stability. Earlier, Toupin and Bernstein [1961] investigated the connexion with wave speeds showing that under conditions of strong ellipticity all waves move with real speeds,[8] while under conditions of semi-strong ellipticity all waves travel with real or zero speeds.[9] Other static examples were given by Zorski [1962], while Hayes [1969] presented additional results in the finite static theory. Hayes [1966] also pointed out that the strong ellipticity condition is not invariant and has suggested an alternative condition free from this defect.

[7] See Truesdell and Noll [1965 §52].
[8] Compare also Hadamard [1903].
[9] Truesdell [1966] examined the relation with longitudinal waves.

Chapter 3

Early Work

The literature on uniqueness in linear elasticity falls roughly into two periods. The first, initiated a little before the appearance of Kirchhoff's classical theorem in 1859, ended in about 1910. The second, starting say in 1950 with the publication of Fichera's paper, still continues to attract interest. In this chapter, we briefly survey the contributions made during the earlier period, devoting particular attention to the work of E. and F. Cosserat.

The first attempt at rigorously establishing uniqueness in elasticity was probably made by Kirchhoff [1850, p. 70] in the course of deriving the equations of plates and shells. The method used in this paper was later extended by Kirchhoff [1859] to the full three-dimensional equilibrium equations of homogeneous isotropic elasticity and became in this form the standard means of establishing uniqueness. For the sake of completeness we now describe the proof; we shall in fact slightly generalise it to include non-homogeneous anisotropic elasticity. Thus, let us suppose that the elasticities possess the symmetries

$$c_{ijkl} = c_{jikl} = c_{ijlk}$$

and further obey the positive definiteness inequality

$$\int_B c_{ijkl} \xi_{ij} \xi_{kl} \, dx > 0 \tag{3.1}$$

for symmetric non-zero tensors ξ_{ij}. Let us assume also that the solution of (2.1.23) and (2.1.24) is non-unique; i.e. there exist two solutions which have the same boundary data and body forces. The difference v_i of these displacements satisfies Eq. (2.1.23) with zero body force subject to homogeneous boundary data (2.1.24). If v_i vanishes identically there is nothing to prove. Thus, we suppose v_i to be non-zero. Multiplication of (2.1.23) by v_i followed by an integration by parts leads directly to the identity

$$\int_B c_{ijkl} \left(\frac{\partial v_i}{\partial x_j} + \frac{\partial v_j}{\partial x_i} \right) \left(\frac{\partial v_k}{\partial x_l} + \frac{\partial v_l}{\partial x_k} \right) dx = 0. \tag{3.2}$$

There are now two possibilities to consider. Either the strains $\left(\dfrac{\partial v_i}{\partial x_j}+\dfrac{\partial v_j}{\partial x_i}\right)$ vanish identically in B or they do not. The second alternative means that (3.1) and (3.2) are in contradiction. Hence the strains are identically zero which means that v_i corresponds to a rigid body displacement, which will be zero whenever the displacement is prescribed on any part of the boundary. This completes the proof of the following theorem.

Theorem 3.1. (Kirchhoff.) *Condition* (3.1) *is sufficient to guarantee that there be at most one classical solution to the mixed boundary value problem of linear elastostatics in bounded regions.*

It should be noted that the above proof is valid if (3.1) is replaced by the negative definiteness condition,

$$\int_B c_{ijkl}\,\xi_{ij}\,\xi_{kl}\,dx<0 \tag{3.3}$$

for symmetric non-zero tensors ξ_{ij}.

Pearson [1893, §1255] has suggested that Kirchhoff obtained the idea for his proof from the discussion of uniqueness contained in Saint-Venant's [1855, especially pp. 293–294] famous memoir on the torsion and bending of beams.[1] The only similarity, however, between the proofs is the consideration of the difference between two solutions satisfying the same data, a device that Saint-Venant credited to Fourier. Further, on remembering that Kirchhoff's first paper appeared in 1850, five years before that by Saint-Venant, it seems hardly possible that Pearson's conjecture is correct. Indeed, it is more likely that Kirchhoff's idea arose directly from the variational procedures he employed in deriving his equations, and so was arrived at quite independently.

In subsequent works, conditions (3.1) and (3.3) are sometimes replaced by the stronger inequalities

$$c_{ijkl}\,\xi_{ij}\,\xi_{kl}>0 \quad \text{or} \quad <0 \tag{3.4}$$

for symmetric non-zero ξ_{ij}, assumed to hold at all points of the body. For isotropic bodies, we have

$$c_{ijkl}=\lambda\,\delta_{ij}\delta_{kl}+\mu(\delta_{il}\delta_{jk}+\delta_{ik}\delta_{jl}),$$

---

[1] Saint-Venant treated uniqueness in connexion with his semi-inverse solution to the problem of bending. His method of proof was almost completely devoid of mathematical reasoning, relying instead upon intuition. Thus, after the formation of the difference of two solutions, he was content to complete his proof with the observation: «On verra que (v_i) seront les déplacement des points d'un prisme que ne serait sollicité que par des forces nulles. Ces déplacements seraient nuls eux-mêmes.» Pearson [1893, §6], while feeling this proof deficient, failed in his criticism of it because the examples he adduced, although appropriate to the non-linear theory, are generally inapplicable to the linear theory. See Truesdell and Noll [1965, pp. 128, 129], John [1958, Chap. IV].

and a necessary and sufficient condition for (3.4) to hold is

$$(3\lambda + 2\mu)\,\mu > 0, \qquad\qquad (3.5)$$

or

$$\mu \neq 0, \quad -1 < \sigma < \tfrac{1}{2}, \qquad\qquad (3.6)$$

which is the classical criterion for uniqueness.

The advantage of Kirchhoff's proof lies in its applicability to boundary problems of mixed type and also in its easy extension to other systems of equations. At the time of its introduction, however, the technique was not readily accepted, since it was believed that there was no adequate justification for supposing the conditions to be physically reasonable. In answer to this objection, Kirchhoff, concerning himself exclusively with (3.1), claimed at first that they were "self-evident"; but later in his Collected Lectures [1877, pp. 392–395][2] he asserted that they followed as a consequence of stability. To support this contention he appealed to a theorem by Lejeune-Dirichlet [1846] which stated that a necessary and sufficient condition for the stability of an equilibrium position in a *discrete* system is that in this position the potential energy must have a minimum value. Kirchhoff erroneously believed that the same criterion would be valid for *unqualified* stability in *continuous* systems (including, therefore, linear elasticity), a mistake persistently repeated in the literature until only recently.[3] Thus, Kirchhoff's argument for the validity of (3.1) must be treated with caution. Nowadays, it is customary merely to state conditions for uniqueness without endeavouring to produce supporting evidence in their favour.

The next major advance in uniqueness theorems came in a series of remarkable papers appearing in the Comptes Rendus between 1898 and 1901. In these the Cosserat brothers, Eugene and François, discussed a method for solving the three-dimensional boundary value problems of homogeneous isotropic elasticity based upon ideas of Picard and Poincaré [1894]. The method, closely associated with the

---

[2] Pearson [1893, §1278] sketched the change. Nevertheless, several authors preferred Kirchhoff's first explanation. Clebsch [1862, p. 68–70] stated (3.1) follows from its "physical meaning", which passed without comment in Saint-Venant's annotated translation [1883, §21]. Kelvin and Tait [1879, §673] remarked merely that the potential energy cannot be permitted "to become negative for any values, positive or negative, of the strain components". They may have had in mind earlier results of Green [1839] on stability. Trefftz [1928] stated it is "physically obvious" that both Lamé constants are positive. Love [1892; 1927, p. 99] was one of the few early writers to accept Kirchhoff's revised claim.

[3] This point is elucidated in the articles by Shield [1965] and Gilbert and Knops [1967].

Laurent expansion of an analytic function, supposed that the displacement is analytic in $\alpha(=(1-2\sigma)^{-1})$ except at certain critical points.

Thus, they[4] assumed the solution of (2.1.23) and (2.1.24) could be represented as

$$\mathbf{u}(\mathbf{x})=\mathbf{u}^0(\mathbf{x})-\alpha\sum_{n=1}^{\infty}\frac{A_n\mathbf{u}^{(n)}(\mathbf{x})}{\alpha-\alpha_n} \tag{3.7}$$

where $\mathbf{u}^0(\mathbf{x})$ is a harmonic vector function satisfying the prescribed data, and $\mathbf{u}^{(n)}(\mathbf{x})$ are elements in the null space of the governing operator, i.e., for $\alpha=\alpha_n$ they are non-zero solutions of Eq. (2.3.2) satisfying homogeneous boundary data. Thus, the $\mathbf{u}^{(n)}(\mathbf{x})$ represent non-unique solutions. It follows easily that for $n \neq m$

$$\int_B \frac{\partial u_i^{(n)}}{\partial x_i}\frac{\partial u_j^{(m)}}{\partial x_j}\,dx=0, \tag{3.8}$$

a relation that serves to determine the constant coefficients A_n in (3.7). Once these coefficients have been found, together with the complete null space, the solution becomes known, in principle, from Eq. (3.7). The original presentation had some obvious deficiencies and was justly criticised by Korn [1908, 1909], but a recent unified study by Mikhlin [1966, 1967] and Mikhlin and Maz'ya [1967] rigorously confirmed many of the results announced by the Cosserats. In particular, for the displacement boundary value problem, these authors established the analyticity of the displacement as a function of α for $|\alpha|<1$, the completeness of the *eigenfunctions* $\mathbf{u}^{(n)}$, the location and multiplicity of the *eigenvalues* α_n, and the validity of expansion (3.7). The traction boundary value problem was likewise treated.

During their investigations, the Cosserats examined the location and structure of the null space in various problems and so obtained improvements upon Kirchhoff's result. It is this aspect of their work that is of present interest.

The Cosserats [1898a] first treated the displacement boundary value problem. Following Kirchhoff, they assumed two solutions to exist and considered their difference. The displacement v_i of the difference solution thus vanished on the entire boundary surface. They next observed that according to earlier work by Borchardt [1873] in the variational derivation of the Navier equations the potential energy for a three-dimensional

---

[4] See E. and F. Cosserat [1898a, 1901a, b]. A comprehensive account of their relevant contributions, including those described in the following pages, was given by Appell [1903, p. 528–532]. Their work was also mentioned by Love [1927, p. 266] and Trefftz [1928, p. 110], although not from the standpoint of uniqueness.

body may be written as

$$V = \tfrac{1}{2} \int_B \left[(\lambda + 2\mu) \frac{\partial v_i}{\partial x_i} \frac{\partial v_j}{\partial x_j} + \frac{\mu}{2} \left(\frac{\partial v_i}{\partial x_j} - \frac{\partial v_j}{\partial x_i} \right) \left(\frac{\partial v_i}{\partial x_j} - \frac{\partial v_j}{\partial x_i} \right) \right] dx. \quad [5] \quad (3.9)$$

Thus provided

$$\mu(\lambda + 2\mu) > 0, \quad (3.11)$$

or equivalently,

$$\mu \neq 0, \quad -\infty < \sigma < \tfrac{1}{2}, \quad 1 < \sigma < \infty, \quad (3.12)$$

the potential energy is always non-zero for displacement fields possessing non-zero dilatation and/or rotations. As in the proof of Kirchhoff's theorem they observed that the potential energy associated with the difference displacement v_i satisfies

$$V(v_i) = 0. \quad (3.13)$$

They now assumed that v_i was not identically zero, otherwise again there was nothing to prove. This led to a contradiction unless

$$\frac{\partial v_j}{\partial x_j} = \frac{\partial v_i}{\partial x_j} - \frac{\partial v_j}{\partial x_i} \equiv 0 \quad \text{in } B, \quad (3.14)$$

showing that the displacement was harmonic. Uniqueness then followed from the uniqueness of the Dirichlet problem in potential theory. We therefore have

Theorem 3.2. (Cosserat.) *Condition* (3.11) *is sufficient to guarantee that there be at most one classical solution to the displacement boundary value problem of homogeneous isotropic elastostatics in bounded regions.*

---

[5] This form of the potential energy was separately discovered by Kelvin [1888] who gave the following calculation: since for arbitrary u_i

$$\left(\frac{\partial u_i}{\partial x_j} + \frac{\partial u_j}{\partial x_i} \right) \left(\frac{\partial u_i}{\partial x_j} + \frac{\partial u_j}{\partial x_i} \right) = \left(\frac{\partial u_i}{\partial x_j} - \frac{\partial u_j}{\partial x_i} \right) \left(\frac{\partial u_i}{\partial x_j} - \frac{\partial u_j}{\partial x_i} \right) + 4 \frac{\partial}{\partial x_j} \left(u_i \frac{\partial u_j}{\partial x_i} \right)$$

$$- 4 \frac{\partial}{\partial x_i} \left(u_i \frac{\partial u_j}{\partial x_j} \right) + 4 \frac{\partial u_i}{\partial x_i} \frac{\partial u_j}{\partial x_j},$$

it follows that

$$V \equiv \tfrac{1}{2} \int_B \left[\lambda \frac{\partial v_i}{\partial x_i} \frac{\partial v_j}{\partial x_j} + \frac{\mu}{2} \left(\frac{\partial v_i}{\partial x_j} + \frac{\partial v_j}{\partial x_i} \right) \left(\frac{\partial v_i}{\partial x_j} + \frac{\partial v_j}{\partial x_i} \right) \right] dx$$

$$= \tfrac{1}{2} \int_B \left[(\lambda + 2\mu) \frac{\partial v_i}{\partial x_i} \frac{\partial v_j}{\partial x_j} + \frac{\mu}{2} \left(\frac{\partial v_i}{\partial x_j} - \frac{\partial v_j}{\partial x_i} \right) \left(\frac{\partial v_i}{\partial x_j} - \frac{\partial v_j}{\partial x_i} \right) \right] dx \quad (3.10)$$

$$+ \mu \int_{\partial B} \left[n_j v_i \frac{\partial v_j}{\partial x_i} - n_i v_i \frac{\partial v_j}{\partial x_j} \right] dS.$$

Clearly, if $v_i \equiv 0$ on ∂B then (3.9) is obtained. Fredholm [1906] supplied an alternative derivation of (3.9), his treatment being later generalized by Fichera [1950]. A discussion of uniqueness in other boundary value problems, making use of (3.10), was given by Bramble and Payne [1962a, §§III, IV]; see also pp. 54, 56.

We remark that condition (3.11) corresponds to the strong ellipticity condition (2.4.13) in the isotropic case (*c.p.* 2.4.16). We prove in the next chapter that the same condition suffices for the uniqueness of the anisotropic problem.

The Cosserats appear to have been the first to announce this result, yet curiously their priority was rarely acknowledged, even by authors writing a few years later. The notable exception is Appell [1903, pp. 528–532], but typical of other writers of this period to whom Theorem 3.2 has occasionally been attributed are Fredholm [1906] and Boggio [1907a]. Of these two authors, Fredholm, in spite of repeating in essence the Cosserats' argument, refered only briefly to them, and Boggio made no reference at all. There was no mention of the result by Love [1927], although the Cosserats' *method of solution* was briefly described; similar remarks apply to Trefftz [1928]. In the course of their investigations into the uniqueness question for unbounded domains, Fichera [1950], followed by Duffin and Noll [1958] and Gurtin and Sternberg [1960], independently rederived the result. Proofs were also given by Ericksen and Toupin [1956] and Hill [1961a]. The arguments of all of these later authors have features resembling, to a larger or smaller extent, those of the original proof. An analogous result in plane elasticity was established by Muskhelisvili [1933, 1953, §40].

When the moduli satisfy

$$\lambda + 2\mu = 0 \tag{3.15}$$

or

$$\mu \neq 0, \quad \sigma = 1 \tag{3.16}$$

a possible solution to the three-dimensional Navier equations is[6]

$$\mathbf{v}(\mathbf{x}) = (\mathbf{x} - \mathbf{x}_0) \times \nabla \varphi + \nabla \psi, \tag{3.17}$$

where $\mathbf{x}$ is the position vector, $\mathbf{x}_0$ is some constant vector, $\varphi(\mathbf{x})$ is a harmonic function, and $\psi(\mathbf{x})$ is an arbitrary function. Elements of this solution, corresponding to $\varphi \equiv 0$, were also given by Ericksen and Toupin [1956] and Hill [1961a]; the latter author also showed that if $\varphi = 0$ in $\bar{B}$ and

$$\nabla \psi = 0 \quad \text{on } \partial B \tag{3.18}$$

then both the associated displacement and surface traction vanish on ∂B. Clearly, this observation implies non-uniqueness of the displacement, traction and mixed boundary value problems for moduli satisfying (3.15) or (3.16), but uniqueness may be trivially recovered in such situations by requiring $u_i \in C^3(B)$ and the dilatation to be harmonic in B.[7]

So far, only sufficient conditions for uniqueness have been established. The question next arises whether or not these conditions are also necessary. For the mixed and traction boundary value problems little

---

[6] See E. and F. Cosserat [1898b].
[7] See E. and F. Cosserat [1898a].

has emerged in this respect until very recently,[8] but in the displacement boundary value problem a complete answer was found by the Cosserats.

In proving that a certain condition is necessary for uniqueness, it must be shown that in all problems of the given class uniqueness implies satisfaction of the stipulated condition. Rather than embark upon this somewhat formidable task, we confront ourselves with the equivalent but simpler program of establishing the inverse of the logical implication. Thus, we may alternatively prove that: *failure of the condition implies non-uniqueness in at least one example.* Usually, the example is selected to be elementary as regards both geometry and material symmetry. This alternative proposition is clearly not invalidated by geometries for which the stated condition fails but uniqueness holds. Indeed, several examples of this type appear later in the Tract. Provided a single example can be demonstrated for which violation of the given condition leads to non-uniqueness, then the necessity of that condition is established. This observation is repeatedly used in what follows.

As a counter-example to uniqueness, the Cosserats [1898c] considered an ellipsoid and produced a class of solutions whose elasticities did not satisfy (3.11) (or 3.12) but whose displacement vanished on[9]

$$\Omega \equiv x_1^2 + a^2 x_2^2 + b^2 x_3^2 - 1 = 0. \tag{3.19}$$

These solutions are of the form

$$u_i = \Omega P_i \tag{3.20}$$

where P_i are polynomials of non-negative degree. For instance, if one takes P_i to be constant, then

$$u_i = \Omega \delta_{i1},$$

and the Navier equations are satisfied provided

$$\mu(\lambda + 2\mu) = -\mu^2(a^2 + b^2) < 0. \tag{3.21}$$

More complicated expressions result for higher order polynomials but for each the corresponding elasticities lie in the range (3.21).

More recently, Ericksen and Toupin [1956] and Hill [1961a] showed non-uniqueness for $\mu = 0$, $\lambda \neq 0$, by observing that any solenoidal displacement satisfying zero data is a solution to the problem.

Thus, when $\mu = 0$, $\lambda \neq 0$, the equilibrium equations (2.3.2) become

$$\lambda \frac{\partial^2 u_j}{\partial x_i \partial x_j} = 0,$$

which clearly are satisfied by imposing the constraint $\dfrac{\partial u_i}{\partial x_i} = 0$, or equivalently: $u_i = e_{ijk} \dfrac{\partial \psi_k}{\partial x_j}$, where e_{ijk} is the permutation symbol and ψ_k an arbitrary vector function of position. To establish non-uniqueness, we may follow Mills [1963], for instance, and take $\psi_k = (0, 0, [f(x_1, x_2, x_3)]^2)$ where $f(\mathbf{x}) = 0$ on any closed surface S. Clearly, u_i vanishes on S but is non-zero inside, thus implying non-uniqueness.

---

[8] See §4.3.4.

[9] Ericksen [1957] obtained the first member of this class for the special case of axisymmetry in n-dimensional regions.

Let us also note that in the absence of body force the displacement is harmonic when $\sigma = \pm \infty$, so that uniqueness follows from standard theorems in potential theory. If $\sigma = \frac{1}{2}$ (incompressibility) a straightforward proof[10] also yields uniqueness.

The preceding remarks together with Theorem 3.2 constitute the proof of the following theorem:

Theorem 3.3. *For bounded regions there is at most one classical solution to the three-dimensional homogeneous isotropic elastostatic displacement boundary value problem if and only if*

$$\mu \neq 0, \quad -\infty \leqq \sigma \leqq \tfrac{1}{2}, \quad 1 < \sigma \leqq \infty. \tag{3.22}$$

We remark again that by "only if" we mean that if the condition is not fulfilled there will be regions for which non-unique solutions exist. For a given domain B there will be at most one solution except for a discrete set of values of σ in the range $\frac{1}{2} < \sigma \leqq 1$.

In view of the significant role that the geometry of the body occupied in the above counter-example to uniqueness, it is worth enquiring what further relaxation can be obtained on the range (3.22) for other particular geometries. In later chapters, detailed analyses are given for the whole space and half space. While most of these results are either new or comparatively recent in origin, the Cosserats, by the beginning of the century, had already extensively investigated the question for homogeneous isotropic bodies with spherical boundaries under various boundary conditions. A description of their conclusions is postponed, as these relate directly to topics treated in Chapter IV (see §4.1.6), but it is worth recording one further result of theirs based upon work by Chree[11] for an ellipsoid. For this body, the Cosserats [1901c] showed that there exist geometries for which the traction boundary value problem for a homogeneous isotropic material has more than one solution provided

$$-\infty < \sigma < -1. \tag{3.23}$$

Thus we may begin to see in what sense the Kirchhoff range is both necessary and sufficient for uniqueness in the traction problem, a matter also further discussed in Chapter IV.

Other proofs of uniqueness in the homogeneous isotropic elastostatic problem are due to Lauricella [1906], Korn [1907], Fredholm [1906] and Boggio [1907a, b]. Most of these authors applied integral equations methods to solve the boundary value problem.

---

[10] The proof, given on p.48 for the traction boundary value problem, may be easily adapted to the present situation.

[11] Chree [1895]. Indeed, since completeness is not required, in principle any solution may yield conditions for non-uniqueness.

Due to the Cosserats, uniqueness in static problems was, during this early period of activity, developed well beyond the original theorem of Kirchhoff. However, in the corresponding problems in elastodynamics, once Neumann [1885][12] had stated his theorem, the topic lay dormant until the present decade when its revitalisation saw the discovery of fairly complete results. Neumann's theorem, strictly analogous to Kirchhoff's, is the following:

Theorem 3.4. (Neumann.) *For bounded regions there is at most one classical solution to the mixed boundary value problem of elastodynamics provided inequality (3.1) is satisfied.*

To prove this theorem, we again consider two solutions (this time of Eq.(2.1.12)) satisfying the same initial and boundary data, and with the same body force. Then the displacement v_i associated with the difference of these two solutions satisfies Eq.(2.1.12) with zero body force and homogeneous initial and boundary conditions. Conservation of energy then gives

$$\int_{B(t)} \rho \frac{\partial v_i}{\partial t} \frac{\partial v_i}{\partial t} dx + \int_{B(t)} c_{ijkl} \frac{\partial v_i}{\partial x_j} \frac{\partial v_k}{\partial x_l} dx = 0 \tag{3.24}$$

in which $B(t)$ denotes integration over the volume B at time t. Let us now assume that v_i is not identically zero. Then, since both integrals appearing in (3.24) are non-negative, (3.24) can be satisfied only if both integrals are identically zero. This implies a zero velocity and therefore a time independent displacement, which being zero initially, is zero for all time. The proof of the theorem is now complete.

Of course, the above argument is not invalidated on replacing (3.1) by the weaker condition

$$\int_{B} c_{ijkl} \xi_{ij} \xi_{kl} dx \geq 0 \tag{3.25}$$

for non-zero symmetric tensors ξ_{ij}, i.e., by a positive semi-definiteness condition on the elasticities. However, we shall see later that such definiteness properties can be removed altogether.

Finally, let us note that Almansi [1907] gave a proof of uniqueness in static homogeneous isotropic elasticity for problems with Cauchy data prescribed over arbitrary portions of the boundary. A generalization of this theorem is given in Chapter VII.

---

[12] Neumann first announced his result in 1859. See the comment by Todhunter and Pearson [1893, §1198].

Chapter 4

Modern Uniqueness Theorems
in Three-Dimensional Elastostatics

As outlined in the previous chapter, uniqueness theorems for the standard boundary value problems of classical linear isotropic elasticity when the elasticities are within the "physical" range, were found by Kirchhoff over a century ago. In addition, the Cosserats, almost forty years later, discovered that in the displacement boundary value problem of isotropic elasticity, uniqueness held for an extended range of values of the elasticities. In general, uniqueness theorems established by early workers made use of an appropriate energy identity. For the most part we continue to use this device in the present chapter in discussing the uniqueness of either weak or classical solutions to various standard boundary value problems in anisotropic and isotropic elasticity. We shall establish conditions sufficient (and in some cases also necessary) for uniqueness in these problems and make special mention of results peculiar to spherical regions.

4.1 The Displacement Boundary Value Problem
for Bounded Regions

We consider three separate cases:

a) A general anisotropic material.
b) An anisotropic material with uniform elasticities.
c) An isotropic material with uniform elasticities.

Theorems proved for case b) are actually stronger than those established in case a) and include the results obtained in Section 4.1.3 for case c).

4.1.1 General Anisotropy.   We begin this section by proving the following theorem:

Theorem 4.1.1. *There is at most one weak solution of the displacement boundary value problem for an anisotropic non-homogeneous material in* *a three-dimensional bounded region B, provided there exists at every point in B a positive constant c_0 such that*

$$c_{ijkl}(\mathbf{x})\,\xi_{ij}\,\xi_{kl} \geq c_0\,\xi_{ij}\,\xi_{ij} \tag{4.1.1}$$

for every tensor ξ_{ij}.

To prove the theorem we assume, as is customary, that two solutions u_i^1, u_i^2 exist and set

$$v_i = u_i^1 - u_i^2.$$

Then from (2.2.5), v_i satisfies

$$\int_B c_{ijkl}\frac{\partial\varphi_i}{\partial x_j}\frac{\partial v_k}{\partial x_l}\,dx = 0 \tag{4.1.2}$$

for every vector function $\varphi_i(\mathbf{x})$ which is continuously differentiable in B and vanishes on ∂B. Since (4.1.2) must hold for limits of C^1 functions, φ_i, in the norm

$$\|\varphi\| = \left[\int_B \frac{\partial\varphi_i}{\partial x_j}\frac{\partial\varphi_i}{\partial x_j}\,dx + \int_B \varphi_i\,\varphi_i\,dx\right]^{\frac{1}{2}}, \tag{4.1.3}$$

it follows by standard arguments that we may take $\varphi_i \equiv v_i$ and thus

$$\int_B c_{ijkl}\frac{\partial v_i}{\partial x_j}\frac{\partial v_k}{\partial x_l}\,dx = 0. \tag{4.1.4}$$

In view of (4.1.1), we conclude that $\dfrac{\partial v_i}{\partial x_j} = 0$ almost everywhere in B and hence that the vector v_i is almost everywhere a constant vector. Since $v_i \in C^0$ in $\bar{B}$ and vanishes on ∂B, it follows that $v_i \equiv 0$ in B, and the theorem is proved. Note that no symmetry assumptions are required on the elasticities, and that instead of (4.1.1) a negative definite condition would serve equally well.

The next two theorems give circumstances in which the positive definiteness criterion may be slightly relaxed but in which uniqueness still holds. Attention is confined, however, to classical solutions.

Theorem 4.1.2. *Let the elasticities satisfy the symmetry condition*

$$c_{ijkl} = c_{kjil}, \tag{4.1.5}$$

the positive semi-definiteness condition

$$c_{ijkl}\,\xi_{ij}\,\xi_{kl} \geq 0, \tag{4.1.6}$$

for all real tensors ξ_{ij}, and the strong-ellipticity condition

$$c_{ijkl}\,\alpha_i\,\alpha_k\,\beta_j\,\beta_l \geq c_1\,\alpha_i\,\alpha_i\,\beta_j\,\beta_j \tag{4.1.7}$$

for some positive constant c_1 and every pair of real vectors α_i, β_i. In addition let the spatial gradients of the elasticities satisfy the boundedness condition

$$\sup_{x \in \bar{B}} \left| \frac{\partial c_{ijkl}}{\partial x_l} \right| < M \tag{4.1.8}$$

for some positive constant M. Then there is at most one classical solution of the displacement boundary value problem for an anisotropic non-homogeneous material in a bounded domain B.

To prove this theorem, we let r denote the distance from some origin inside B and we let a denote the radius of a circumscribing sphere with centre at the origin. For any positive constant p we then have

$$\int_B (a^p - r^p) c_{ijkl} \frac{\partial v_i}{\partial x_j} \frac{\partial v_k}{\partial x_l} dx = \frac{1}{2} \int_B \frac{\partial}{\partial x_l} \left\{ c_{ijkl} \frac{\partial}{\partial x_j} (a^p - r^p) \right\} v_i v_k dx, \tag{4.1.9}$$

where as before v_i represents the difference of two solutions to the problem. Here we have used integration by parts and the fact that v_i vanishes on ∂B. Eq. (4.1.9) may be expanded into the form

$$\int_B (a^p - r^p) c_{ijkl} \frac{\partial v_i}{\partial x_j} \frac{\partial v_k}{\partial x_l} dx + \frac{p}{2} \left\{ \int_B c_{ijkj} r^{p-2} v_i v_k dx + \right.$$

$$\left. + (p-2) \int_B r^{p-4} c_{ijkl} x_j x_l v_i v_k dx + \int_B r^{p-2} \frac{\partial c_{ijkl}}{\partial x_l} x_j v_i v_k dx \right\} = 0. \tag{4.1.10}$$

The first term on the left of (4.1.10) is non-negative and moreover

$$\int_B r^{p-2} \frac{\partial c_{ijkl}}{\partial x_l} x_j v_i v_k dx \geqq -k_1 \int_B r^{p-1} v_i v_i dx \geqq -k_1 a \int_B r^{p-2} v_i v_i dx,$$

$$\int_B r^{p-2} c_{ijkj} v_i v_k dx = \int_B r^{p-2} c_{ilkm} v_i v_k \delta_l^j \delta_m^j dx \geqq 3c_1 \int_B r^{p-2} v_i v_i dx,$$

for computable constant k_1 independent of p. Thus, dropping the first term on the left of (4.1.10), and making use of (4.1.7) we obtain

$$[(p+1) c_1 - k_1 a] \int_B r^{p-2} v_i v_i dx \leqq 0.$$

By choosing p sufficiently large, the term in square brackets can be made positive which implies that v_i vanish identically in B. The theorem is thus proved.

Note that (4.1.5) is not the major symmetry condition

$$c_{ijkl} = c_{klij} \tag{4.1.11}$$

and is in fact not satisfied for classical isotropic elasticity.

The third theorem of this section deals with *star-shaped* domains. Let us assume that at each point of the boundary ∂B we have

$$h \equiv x_k n_k \geq h_0 > 0, \tag{4.1.12}$$

where n_k is the unit outward normal on ∂B and h_0 is some positive constant. For the purposes of the theorem, we must further suppose that the elasticities are analytic in the region B of the elastic material and are extendable as analytic functions into a neighbourhood of some segment Σ' of ∂B of non-zero measure. We can then prove the following theorem:

Theorem 4.1.3. *There is at most one solution of the displacement boundary value problem for an anisotropic non-homogeneous material in a bounded star-shaped domain B, provided the elasticities are analytic and extendable in the sense just described, satisfy the symmetry and strong ellipticity conditions (4.1.11) and (4.1.7), respectively, and the condition*

$$b_{ijkl}(\mathbf{x}) \, \xi_{ij} \, \xi_{kl} \equiv \left[x_m \frac{\partial c_{ijkl}}{\partial x_m} + \beta \, c_{ijkl} \right] \xi_{ij} \, \xi_{kl} \geq 0, \tag{4.1.13}$$

for some constant β and every tensor ξ_{ij}.

Because of strong ellipticity, it is well known (see, e.g., Bers and Schechter [1964, p. 136]) that the solution to the problem in question is analytic and therefore is a solution in the classical sense.

As usual, we suppose that there are two solutions u_i^1 and u_i^2 and we set $v_i = u_i^1 - u_i^2$. Then upon integrating by parts and using the fact that v_i vanishes on ∂B, we have

$$
\begin{aligned}
0 &= \int_B x_m \frac{\partial v_i}{\partial x_m} \frac{\partial}{\partial x_j} \left[c_{ijkl} \frac{\partial v_k}{\partial x_l} \right] dx \\
&= \tfrac{1}{2} \int_{\partial B} h \, c_{ijkl} \frac{\partial v_i}{\partial x_p} \frac{\partial v_k}{\partial x_q} n_p \, n_q \, n_j \, n_l \, dS \\
&\quad + \tfrac{1}{2} \int_B \left[b_{ijkl} + (1-\beta) \, c_{ijkl} \right] \frac{\partial v_i}{\partial x_j} \frac{\partial v_k}{\partial x_l} dx.
\end{aligned}
\tag{4.1.14}
$$

But from the ordinary Green's identity it follows that

$$\int_B c_{ijkl} \frac{\partial v_i}{\partial x_j} \frac{\partial v_k}{\partial x_l} dx = 0, \tag{4.1.15}$$

and thus in view of (4.1.7), (4.1.12), and (4.1.13) we may conclude that

$$h_0 \, c_1 \int_{\partial B} n_p \frac{\partial v_i}{\partial x_p} n_q \frac{\partial v_i}{\partial x_q} dS \leq 0. \tag{4.1.16}$$

We see therefore that not only v_i but also its normal derivative vanishes on ∂B. It then follows from Holmgren's theorem[1] that v_i vanishes identically in B, and the theorem is proved.[2]

It should be observed that the theorem is still valid in situations where, for instance, the elasticities are bounded and analytic at every point except at the origin, the result now following by continuity. This observation is important for a spherical domain possessing transverse isotropy such that the elasticities are independent of the radial variable. Holmgren's theorem actually shows that if the conditions of Theorem 4.1.3 are met only in some finite neighbourhood N of the boundary, then the solution of the displacement boundary value problem is unique in N but not necessarily unique in $B - \overline{N}$. This is readily demonstrated on assuming B to be a sphere and in $B - \overline{N}$ taking

$$c_{ijkl} = \mu [\delta_{ik} \delta_{jl} + \delta_{il} \delta_{jk} - 2\delta_{ij} \delta_{kl}], \qquad (4.1.17)$$

(i.e., $\sigma = 1$), but in $\overline{N}$ letting the elasticities satisfy (4.1.11), (4.1.7) and (4.1.13). Clearly, if φ is any smooth function vanishing with its third derivatives on $\partial (B - N)$ then

$$v_i = \begin{cases} \dfrac{\partial \varphi}{\partial x_i} & \text{in } B - \overline{N} \\ 0 & \text{in } \overline{N} \end{cases} \qquad (4.1.18)$$

is of class C^2 in B and is a solution of the governing equation. We see therefore that in this case the solution of the displacement boundary value problem is not unique in all of B.

4.1.2 A Homogeneous Anisotropic Material. The elasticities are now assumed to be *uniform*, a condition which enables uniqueness to be established under the single assumption of strong ellipticity. In the next section we prove the necessity of this condition.

Theorem 4.1.4. *There is at most one weak solution of the displacement boundary value problem for an anisotropic material in a bounded three-dimensional region provided the elasticities are uniform and satisfy the strong ellipticity condition*

$$c_{ijkl} \alpha_i \alpha_k \beta_j \beta_l \geq c_1 \alpha_i \alpha_i \beta_j \beta_j \qquad (4.1.7)$$

for some positive constant c_1 and every pair of real vectors α_i, β_i.

---

[1] Holmgren's theorem states: The Cauchy problem for a linear system of differential equations with analytic coefficients and with Cauchy data prescribed on an analytic non-characteristic initial surface S has at most one solution in a neighbourhood of S. Compare John [1964, p. 47] who, besides giving further references, also mentions that S need not be analytic and that it is sufficient to assume the solution is of class C^m on one side of S and of class C^{m-1} on S. Here, m is the order of the differential equation.

[2] The governing equations are strongly-elliptic and therefore elliptic and so the characteristic surfaces are imaginary.

For classical solutions, this theorem has been obtained by Browder [1954], Morrey [1954], Zorski [1962] and in a modified form by Hayes [1966].

The proof given here employs the method of van Hove [1947]. We observe first, using the same notation and argument of the previous theorems, that

$$\int_B c_{ijkl} \frac{\partial v_i}{\partial x_j} \frac{\partial v_k}{\partial x_l} dx = 0. \tag{4.1.19}$$

Secondly, since v_i vanishes on ∂B, it may be extended to the whole space E_3 by putting $v_i \equiv 0$ in $E_3 - \bar{B}$. From the assumed smoothness of v_i, it follows that $v_i(\mathbf{x})$ and $\partial v_i(\mathbf{x})/\partial x_j$ are absolutely and square integrable over E_3. Hence, they possess Fourier transforms η_i and η_{ij}, respectively given by

$$\eta_i(\mathbf{y}) = (2\pi)^{-\frac{3}{2}} \int_{E_3} e^{ix_k y_k} v_i(\mathbf{x}) dx, \tag{4.1.20}$$

$$\eta_{ij}(\mathbf{y}) = (2\pi)^{-\frac{3}{2}} \int_{E_3} e^{ix_k y_k} \frac{\partial v_i}{\partial x_j}(\mathbf{x}) dx. \tag{4.1.21}$$

An integration by parts in the last formula immediately yields

$$\eta_{ij}(\mathbf{y}) = -i y_j \eta_i(\mathbf{y}). \tag{4.1.22}$$

A well-known theorem in Fourier analysis (see e.g., Goldberg [1961]), then gives

$$\int_{E_3} \frac{\partial v_i}{\partial x_j}(\mathbf{x}) \frac{\partial v_k}{\partial x_l}(\mathbf{x}) dx = \int_{E_3} \bar{\eta}_{ij}(\mathbf{x}) \eta_{kl}(\mathbf{x}) dx, \tag{4.1.23}$$

the bar denoting the complex conjugate. Thus after an insertion of (4.1.22) into (4.1.23) we get

$$\int_B \frac{\partial v_i}{\partial x_j}(\mathbf{x}) \frac{\partial v_k}{\partial x_l}(\mathbf{x}) dx = \int_{E_3} x_j x_l \operatorname{Re}[\eta_i \eta_k] dx. \tag{4.1.24}$$

Multiplication of (4.1.24) by c_{ijkl} and use of (4.1.7) then leads to

$$\int_B c_{ijkl} \frac{\partial v_i}{\partial x_j} \frac{\partial v_k}{\partial x_l} dx = \operatorname{Re} \int_{E_3} c_{ijkl} x_j x_l \bar{\eta}_i \eta_k \, dx$$

$$\geqq c_1 \int_{E_3} x_j x_j \bar{\eta}_i \eta_i \, dx, \tag{4.1.25}$$

$$= c_1 \int_B \frac{\partial v_i}{\partial x_j} \frac{\partial v_i}{\partial x_j} dx.$$

Since the last term on the right is clearly non-negative, we conclude with the help of (4.1.19) that $\partial v_i/\partial x_j$ vanishes almost everywhere in B.

The argument of the previous theorem then shows that v_i is identically zero in B and so the present theorem is proved.[3] Note that, as in the previous theorem, the elasticities are not required to be symmetric.

4.1.3 A Homogeneous Isotropic Material.

In terms of the shear modulus μ and Poisson's ratio σ, the elasticities in the isotropic case may be expressed as

$$c_{ijkl} = \mu[\delta_{ik}\delta_{jl} + \delta_{il}\delta_{jk} + 2\sigma(1-2\sigma)^{-1}\delta_{ij}\delta_{kl}], \qquad (4.1.27)$$

where δ_{ij} is the Kronecker delta. (We remark that a composite isotropic medium can also be treated when μ and σ have different constant values for each constituent material.) Condition (4.1.7) is then satisfied provided $\mu > 0$, $\lambda + 2\mu > 0$ or equivalently

$$\mu > 0, \quad -\infty < \sigma < \tfrac{1}{2}, \quad 1 < \sigma < \infty. \qquad (4.1.28)$$

As indicated earlier, we can equally well impose a negative definiteness condition on the elasticities corresponding to the inequality (2.4.14). Moreover, we may also prove that in the incompressible case ($\sigma = \tfrac{1}{2}$) uniqueness continues to hold (the proof is deferred until later[4]) and we have earlier shown that there is uniqueness when Poisson's ratio has the values $\sigma = \pm\infty$. Thus, in homogeneous isotropic elasticity the displacement problem for bounded regions is, in fact, unique for all moduli in the range

$$\mu \neq 0, \quad -\infty \leqq \sigma \leqq \tfrac{1}{2}, \quad 1 < \sigma \leqq \infty. \qquad (4.1.29)$$

The Cosserats, as already described in Chapter 3, demonstrated by means of a definite example involving an ellipsoid that there can be non-uniqueness for all values of the moduli not in the range (4.1.29). Thus, this range of values represents necessary and sufficient conditions for uniqueness, a result established in Theorem 3.3.[5]

On bearing in mind the remarks on p. 29 concerning conditions necessary for uniqueness, we see that because isotropy is a special case of anisotropy the Cosserats' counter example of the ellipsoid also establishes the necessity of strong ellipticity for uniqueness in the anisotropic problem of Theorem 4.1.4. Additional comments on the status of strong ellipticity are given in the following section.

---

[3] Following the method adopted by Hayes [1966], the proof may be completed differently. From (4.1.25) it follows that $|\eta_i| = 0$ a.e., and then the relation $\int_B v_i v_i \, dx = \int_{E_3} \bar{n}_i \eta_i \, dx$ implies $v_i \equiv 0$ a.e. But v_i is continuous and so vanishes identically everywhere in B.

[4] See p. 48.

[5] Recent proofs of Theorem 3.3 are due to Ericksen and Toupin [1956] and Hill [1961a]. Hill based his proof on an extremum principle, described in a general setting in Hill [1961b].

4.1.4 The Implication of Strong Ellipticity for Uniqueness.   In Section 4.1.2, we proved that for uniform elasticities the strong ellipticity of the governing equations is sufficient for uniqueness in the displacement boundary value problem for bounded regions. There have been several conjectures in the literature that the same conclusion must hold in the non-homogeneous problem. That this is incorrect may be seen from the following example constructed by Edelstein and Fosdick [1968]. Let (r, θ, φ) denote the spherical polar coordinates and let n, m be arbitrary positive integers. Consider the spherical shell $n\pi \leq r \leq (n+m)\pi$ composed of non-homogeneous isotropic elastic material and subjected to a spherically symmetric displacement field. The isotropic elasticities are functions of the radial coordinate alone and are given by

$$\lambda + 2\mu = \frac{1}{r^2}, \quad \mu = \frac{1}{4}\left\{\frac{3}{r^2} + \ln\left[\frac{(n+m+1)\pi}{r}\right]\right\}, \qquad (4.1.30)$$

so that clearly $\mu(\lambda + 2\mu) > 0$, and the strong ellipticity condition is satisfied. It may be easily verified, however, that the displacement field

$$(u_r, u_\theta, u_\varphi) = (\sin r, 0, 0) \qquad (4.1.31)$$

vanishes on both surfaces of the shell and inside satisfies the Navier equation

$$\frac{d}{dr}\left[(\lambda + 2\mu)\left(\frac{du}{dr} + 2\frac{u}{r}\right)\right] - \frac{4u}{r}\frac{d\mu}{dr} = 0. \qquad (4.1.32)$$

This demonstrates that non-uniqueness is possible under the assumption of strong ellipticity.

Care, however, is needed in interpreting this result, since expression (3.9), crucial to the proof of Theorem 3.2, clearly remains valid with λ allowed to vary but with μ held fixed.

Observe further that from (4.1.30) we also obtain

$$3\lambda + 2\mu = -\ln\left[(n+m+1)\frac{\pi}{r}\right] < 0, \qquad (4.1.33)$$

which might lead one to suspect that the condition of positive-definiteness or negative-definiteness proved sufficient for uniqueness in Theorem 4.1.1, might also be necessary for uniqueness in non-homogeneous bodies. However, to establish necessity in this case it must be demonstrated that failure of the definiteness condition at only one point of B leads to failure of uniqueness. That this is unlikely is further indicated by examples briefly considered in the next section.

Let us remark again that for uniform elasticities the condition of strong ellipticity is both necessary and sufficient for uniqueness of the displacement boundary value problem in bounded regions. This does not mean that failure of the strong ellipticity condition results in non-uniqueness in *every* such boundary value problem. Indeed, by suitably restricting the geometry of problems being considered, the strong ellipticity condition may be ignored entirely in establishing necessary conditions for uniqueness. For instance, in Section 4.1.6, we describe how Bramble and Payne, using a complete solution, establish for an isotropic sphere that there is uniqueness in the displacement problem if, and only if, Poisson's ratio does not assume certain discrete values. Naturally, all the exceptional values of Poisson's ratio lie outside the range of values for which the governing equations are strongly elliptic, but more importantly, there are also non-exceptional values of Poisson's ratio lying outside the same range for which there is still uniqueness. Thus, the example of Bramble and Payne illustrates the known fact that strong ellipticity may fail but uniqueness continue to hold.

A similar example is provided by the half-space which is explicitly treated in Chapter 6. Several more instances of this nature are recorded without comment in the following pages.

Another way of showing that strong ellipticity is not always necessary for uniqueness is by restricting the material symmetry of the elastic body. Hayes [1963] has given several counter-examples and listed all the various possibilities when uniqueness holds and does not hold for a transversely isotropic material. In fact, Hayes [1966] has remarked that strong ellipticity is too strong a condition for uniqueness in the general homogeneous anisotropic displacement problem and has proposed the weaker criterion of *moderately strong ellipticity*. This is defined as

a) $c_{ijkl}\xi_i\xi_k\eta_j\eta_l \geq 0$ for all ξ_i, η_i,

b) for $\eta_i \neq 0$,   (4.1.34)

$$c_{ijkl}\xi_i\xi_k\eta_j\eta_l = 0 \Rightarrow \xi_i = 0.$$

Under this condition, Hayes [1966] established uniqueness by means of a proof following closely that of Theorem 4.1.4.

4.1.5 The Non-Homogeneous Isotropic Material with no Definiteness Assumptions on the Elasticities.   We briefly describe two examples of the displacement boundary value problem for non-homogeneous isotropic elasticity, for which uniqueness either does, or does not, hold but for which the definiteness conditions

$$3\lambda + 2\mu > 0, \quad \mu \neq 0,   \qquad (4.1.35)$$

are violated in some way. In addition to Theorems 4.1.2, 4.1.3, these examples provide further information on the status of the definiteness criterion of Theorem 4.1.1, indicating that probably it is too stringent for uniqueness in the displacement boundary value problem of non-homogeneous elasticity. Of course, the results of the previous section have shown that strong ellipticity by itself is not stringent enough.

First of all, it is obvious that if $\mu \equiv 0$ then

$$u_i = e_{ijk} \frac{\partial \psi_k}{\partial x_j} \tag{4.1.36}$$

continues to satisfy the Navier equations for any function ψ_k, in spite of non-homogeneity. Clearly then for $\mu \equiv 0$ the solutions of any of the standard boundary value problems are non-unique.

Again, at the beginning of the previous section it was shown that for $(3\lambda + 2\mu) < 0$, $\mu \neq 0$, there is at least one example demonstrating failure of uniqueness. On the other hand, when

$$3\lambda + 2\mu = 0, \qquad \mu \neq 0,$$

i.e., $\hspace{12em}$ (4.1.37)

$$\sigma = -1, \qquad \mu \neq 0,$$

hold at all points of the body, Edelstein and Fosdick[6] [1968] have proved that *for a non-homogeneous elastic body occupying a smooth bounded region there is uniqueness in the displacement and mixed[7] problems, while in the traction problem there is uniqueness to within the additive displacement field*

$$u_i = \tfrac{1}{2} c_i x_m x_m - c_m x_m x_i + c x_i + A_{ij} x_j + a_i \tag{4.1.38}$$

where c_i, c, $A_{ij} = -A_{ji}$ *and* a_i *are arbitrary constants*[8].

Other examples in which (4.1.35) is violated may be constructed, for instance, by following the pattern described in the previous section for the problems concerning the elastic shell.

4.1.6 The Displacement Boundary Value Problem for a Homogeneous Isotropic Sphere.
We now make some remarks on the displacement boundary value problem for the homogeneous isotropic elastic sphere. It follows from the work of the Cosserats [1898b] and Bramble and Payne [1961a][9] that if B is the interior of a sphere, then there exist non-

---

[6] Actually, they gave the proof only for $\mu > 0$, but it is equally true for $\mu \neq 0$.

[7] I.e., problem (a) of Section 4.4.

[8] Elements of expression (4.1.38) have previously been recorded by Bramble and Payne [1961a, 1962b]. Specialisations of (4.1.38) for a sphere were given by E. and F. Cosserat [1901b], and Boggio [1907b, p. 448]. Earlier versions of the theorem for the traction problem in a *homogeneous* body were presented by Ericksen and Toupin [1956] and Hill [1961a].

[9] See also Love [1927, p. 266] and Bramble [1960].

unique solutions whenever the values of Poisson's ratio are either $\sigma = 1$ or

$$\sigma_n = \tfrac{1}{2}(1+3n)(1+2n)^{-1}, \qquad n = 1, 2, \ldots. \tag{4.1.39}$$

All the values of σ_n lie between $\tfrac{2}{3}$ and $\tfrac{3}{4}$ and so are included in the range predicted by Theorem 3.3. Indeed, the results of Theorem 3.3 may be further refined in the example under consideration. For, the analysis of Bramble and Payne gives a complete solution to the problem and hence it follows that the exceptional values are both necessary and sufficient for non-uniqueness.

If B is exterior to a spherical surface and it is assumed that the displacement field obeys

$$u_i = O(r^{-1}), \qquad r \to \infty, \tag{4.1.40}$$

where r is the distance from some origin in the complement of B, the exceptional values of Poisson's ratio are given by[10] $\sigma = 1$ and

$$\sigma_m = \tfrac{1}{2}(5+3m)(3+2m)^{-1}, \qquad m = 0, 1, 2, \ldots. \tag{4.1.41}$$

All the values of σ_m lie between $\tfrac{3}{4}$ and $\tfrac{5}{6}$.

Because the half-space may be regarded as the limit of regions interior and exterior to a sphere, we may conjecture that exceptional values of σ in the half-space problem lie at the intersection of those for the interior and exterior sphere problems. Hence, from (4.1.39) and (4.1.41) these values should be $\sigma = \tfrac{3}{4}, 1$. This conjecture is, in fact, true and is rigorously established in Chapter 6.

Finally, we repeat the Cosserats'[11] observation that $\sigma = 1$ is no longer exceptional if the dilatation $\partial u_j / \partial x_j$ is required to be harmonic, a property it possesses for all other values of σ.

4.1.7 Fichera's Maximum Principle.

Many of the uniqueness results established so far have relied upon energy arguments which always carry the implied assumption that the strain energy remains bounded. Although for more general problems, other means for obtaining uniqueness are yet to be devised, in the displacement boundary value problem for a *homogeneous isotropic* elastic body occupying a bounded (three-dimensional) region B we may abandon the hypothesis of a bounded strain energy and use instead a maximum principle due to Fichera [1961]. This principle states that provided the region B has a sufficiently regular surface ∂B then each solution of

$$\mu \, \Delta u_i + (\lambda + \mu) \frac{\partial^2 u_j}{\partial x_i \, \partial x_j} = 0 \quad \text{in } B \tag{4.1.42}$$

---

[10] See E. and F. Cosserat [1901d] and Bramble and Payne [1961a].
[11] See E. and F. Cosserat [1898a] and p. 28.

in the class $C^0(B \cup \partial B) \cup C^2(B)$ and subject to

$$-\infty < \sigma < \tfrac{1}{2}, \quad 1 < \sigma < \infty, \ \mu \neq 0, \tag{4.1.43}$$

satisfies the inequality

$$\max_{B \cup \partial B} |\mathbf{u}| \leq H \max_{\partial B} |\mathbf{u}| \tag{4.1.44}$$

where H is a positive constant depending only on the geometry of B and the constant $\alpha = (1 - 2\sigma)^{-1}$.

Clearly, we may use (4.1.44) to prove uniqueness within the given class of solutions.

4.2 Exterior Domains

All the theorems thus far discussed are restricted to bounded domains, and their extension to unbounded domains has still to be considered. The simplest type of unbounded domain, namely the whole space, is explicitly treated in Chapter 6. Also treated in that chapter are a number of other uniqueness theorems for the half-space. Here, we survey another type of unbounded domain in which the region lies external to a finite number of non-intersecting closed regular surfaces. For such exterior regions, the previous theorems remain valid, provided that, in addition, we require the limits of the pertinent boundary integrals over spheres of radius r to tend to zero as r tends to infinity (where r is the distance measured from some origin located in the region). Sufficient conditions for this requirement are easily given. For example, in either the classical version of Kirchhoff's theorem or Cosserats' theorem (see Theorem 3.2) it suffices that the displacement and stress components possess the asymptotic behaviour

$$u_i = u_i(\infty) + O(r^{-1}), \quad \sigma_{ij} = \sigma_{ij}(\infty) + O(r^{-2}), \quad r \to \infty \tag{4.2.1}$$

where $u_i(\infty)$ and $\sigma_{ij}(\infty)$ are prescribed functions defined in the neighbourhood of infinity, and the standard order of magnitude notation is employed. However, as Gurtin and Sternberg[12] [1961a] have pointed out, it is artificial to prescribe a priori the conditions (4.2.1). The rate at which the components of displacement and stress approach their asymptotic values at infinity is a property of the solution to the problem, and is not to be imposed at the outset. Moreover, the prescription of asymptotic values for both the displacement and stress is unnecessary—if indeed compatible.

---

[12] See also Turteltaub and Sternberg [1968].

Thus, it becomes desirable to relax conditions (4.2.1) to either

$$u_i = u_i(\infty) + o(1) \qquad \text{as } r \to \infty \qquad\qquad (4.2.2)$$

or

$$\sigma_{ij} = \sigma_{ij}(\infty) + o(1) \qquad \text{as } r \to \infty. \qquad\qquad (4.2.3)$$

Little advance has been made beyond this point for anisotropic media, but in the homogeneous isotropic case, Fichera [1950], relying upon results of Picone [1936] for biharmonic functions, showed that in the absence of body force, and provided $\sigma \neq 1$, $\mu \neq 0$, condition (4.2.2) implies (4.2.1). He was thus able to generalise Kirchhoff's theorem to exterior domains under the extra assumption that the displacement tends uniformly to its prescribed value at infinity. Under the same assumption he also generalised the Cosserat theorem and proved uniqueness holds in the exterior displacement problem for values of the isotropic elasticities in the extended range (3.11). Duffin and Noll [1958], Gurtin and Stern-berg [1960], and Finn and Chang [1961], each using different means based upon estimates due to Finn and Noll [1957], recovered Fichera's result for the displacement problem. Later, by employing a series expansion in terms of solid harmonics, Gurtin and Sternberg [1961a] returned to the problem and, besides rederiving Fichera's results, also proved that the asymptotic behaviour (4.2.1) is alternatively ensured by the prescription (4.2.3) of the stress components at infinity. Their proof is carried through only for moduli in the "physical" range

$$\mu > 0, \qquad -1 < \sigma < \tfrac{1}{2}$$

but is in fact true for all values except those given by $\mu = 0$, $\sigma = 1$. (This observation is important when considering the generalisation of sub-sequent results to exterior domains.) Thus, Theorems 3.1 (in its classical form) and 3.2 may be extended to exterior domains under the condition that the stress at infinity behaves like (4.2.3). It should be noted, however, that in this case the uniqueness of the displacement boundary value problem requires not only the conditions of Theorem 3.2 and (4.2.3) to be met, but also that the resultant traction and moment of tractions be prescribed over the interior surface. Gurtin and Sternberg [1961a] give counter-examples showing that uniqueness fails on omission of these extra requirements.

4.3 The Traction Boundary Value Problem

In this section we examine the question of uniqueness of weak solutions for bounded three-dimensional domains when the traction is prescribed at every point of the bounding surface. In particular, we treat:

a) A general anisotropic material.

b) An isotropic homogeneous material.

We shall see that the sufficient conditions obtained for uniqueness in b) cannot be obtained from specialising those in a).

4.3.1 General Anisotropy.   The following theorem is established.

Theorem 4.3.1. *In the traction boundary value problem for a non-homogeneous anisotropic elastic body in a bounded three-dimensional region any two weak solutions of (2.1.23) can differ only by a rigid body displacement provided there exists at every point in B a positive constant c_0 such that*

$$c_{ijkl}\, \xi_{ij}\, \xi_{kl} \geq c_0 (\xi_{ij} + \xi_{ji})(\xi_{ij} + \xi_{ji}), \qquad (4.3.1)$$

for every tensor ξ_{ij}.

We begin the proof of this theorem by noting that the method adopted in the proof of Theorem 4.1.1 for the corresponding displacement boundary value problem is valid in the present context and leads to the conclusion that $\partial v_i/\partial x_j + \partial v_j/\partial x_i \equiv 0$ almost everywhere in B. This implies that v_i is almost everywhere of the form

$$v_i = \alpha_i + e_{ijk}\, \beta_j\, x_k \qquad (4.3.2)$$

where α_i and β_i are constants and e_{ijk} is the permutation symbol. By the assumed continuity, v_i must have the form (4.3.2) at every point of B, and the theorem is proved. Note that the elasticities need not satisfy any symmetry requirements. Also observe that under the normalisation (2.1.21) the arbitrary rigid body displacement is fixed to be identically zero, and hence that in these circumstances the solution is unique.

The proof of the next theorem is not given since it follows in an obvious manner from the preceding one.

Theorem 4.3.2. *In the traction boundary value problem for a non-homogeneous anisotropic medium occupying a bounded region any two weak solutions of (2.1.23) can differ only by a rigid body translation, provided there exists at each point in B a positive constant c_1 such that*

$$c_{ijkl}(\mathbf{x})\, \xi_{ij}\, \xi_{kl} \geq c_1\, \xi_{ij}\, \xi_{ij}$$

for every tensor ξ_{ij}.

4.3.2 A Homogeneous Isotropic Material.
If the elasticities are expressed in their isotropic form (4.1.27), Theorem (4.3.1) gives the classical conditions (3.5, 3.6) for uniqueness in the traction boundary value problem for bounded regions. However, for that broad class of problems

concerned with homogeneous isotropic media occupying star-shaped regions, this range of admissible values has been extended by Bramble and Payne [1962a] with the following theorem for classical solutions:

Theorem 4.3.3. *In a bounded three-dimensional isotropic homogeneous elastic body the traction boundary value problem has at most one stress field, provided the moduli satisfy* $\mu \neq 0$,

$$-1 < \sigma < 1, \tag{4.3.3}$$

and at each point of the boundary, the condition

$$h \equiv x_k n_k \geq h_0 > 0 \tag{4.1.12}$$

holds, i.e., *the region is star-shaped with respect to the origin.*

In order to prove the theorem, it suffices to show that any vector v_i satisfying

$$\Delta v_i + \alpha\, \partial^2 v_j / \partial x_j\, \partial x_i = 0 \tag{4.3.4}$$

in B, and

$$n_j \sigma_{ij} = 0 \tag{4.3.5}$$

on ∂B implies $\sigma_{ij} \equiv 0$ in B. Here

$$\alpha = (1 - 2\sigma)^{-1}, \tag{4.3.6}$$

and the n_j are the cartesian components of the unit outward normal on ∂B. We observe first that from Green's identity it follows that the potential energy $V(v_i)$ satisfies

$$V(v_i) = \int_B \left[\frac{\partial v_i}{\partial x_j} \left(\frac{\partial v_i}{\partial x_j} + \frac{\partial v_j}{\partial x_i} \right) + (\alpha - 1) \frac{\partial v_j}{\partial x_j} \frac{\partial v_i}{\partial x_i} \right] dx = 0. \tag{4.3.7}$$

Integration by parts together with (4.3.5) and (4.3.7) next shows that

$$\int_B x_k \frac{\partial v_i}{\partial x_k} \left\{ \Delta v_i + \alpha \frac{\partial^2 v_j}{\partial x_j \partial x_i} \right\} dx$$

$$= -\frac{1}{4} \int_{\partial B} x_k n_k \left[\left(\frac{\partial v_i}{\partial x_j} + \frac{\partial v_j}{\partial x_i} \right) \left(\frac{\partial v_i}{\partial x_j} + \frac{\partial v_j}{\partial x_i} \right) + 2(\alpha - 1) \frac{\partial v_j}{\partial x_j} \frac{\partial v_i}{\partial x_i} \right] dS, \tag{4.3.8}$$

which, because of (4.3.4), leads to

$$\int_{\partial B} x_k n_k \left[\left(\frac{\partial v_i}{\partial x_j} + \frac{\partial v_j}{\partial x_i} \right) \left(\frac{\partial v_i}{\partial x_j} + \frac{\partial v_j}{\partial x_i} \right) + 2(\alpha - 1) \frac{\partial v_j}{\partial x_j} \frac{\partial v_i}{\partial x_i} \right] dS = 0. \tag{4.3.9}$$

However, in view of the stress-strain relation (2.1.8),[13] we may rewrite the condition $\sigma_{ij} n_i n_j = 0$ on ∂B as

$$(\alpha - 1)\frac{\partial v_j}{\partial x_j} = -2\frac{\partial v_j}{\partial x_k} n_j n_k \quad \text{on } \partial B, \quad (4.3.10)$$

and then (4.3.9) becomes for $\alpha \neq 1$,

$$\int_{\partial B} x_k n_k \left[\left(\frac{\partial v_i}{\partial x_j} + \frac{\partial v_j}{\partial x_i}\right)\left(\frac{\partial v_i}{\partial x_j} + \frac{\partial v_j}{\partial x_i}\right) - 8(\alpha - 1)^{-1}\left(\frac{\partial v_j}{\partial x_l} n_j n_l\right)^2\right] dS = 0. \quad (4.3.11)$$

Now suppose $\beta^2 \equiv (\alpha + 1)/(\alpha - 1) \geq 0$. Then (4.3.11) may be expressed as

$$\int_{\partial B} x_k n_k \left[\left(\frac{\partial v_i}{\partial x_j} + \frac{\partial v_j}{\partial x_i}\right) - 2(1 \pm \beta)\frac{\partial v_m}{\partial x_l} n_m n_l n_i n_j\right]$$
$$\cdot \left[\left(\frac{\partial v_i}{\partial x_j} + \frac{\partial v_j}{\partial x_i}\right) - 2(1 \pm \beta)\frac{\partial v_m}{\partial x_l} n_m n_l n_i n_j\right] dS = 0, \quad (4.3.12)$$

where either both plus signs or both minus signs in front of β are to hold. The domain B is star-shaped; therefore (4.1.12) holds on ∂B and so equation (4.3.12) implies that

and

$$\frac{\partial v_i}{\partial x_j} + \frac{\partial v_j}{\partial x_i} - 2(1 + \beta)\frac{\partial v_k}{\partial x_l} n_k n_l n_i n_j = 0 \quad \text{on } \partial B, \quad (4.3.13)$$

$$\frac{\partial v_i}{\partial x_j} + \frac{\partial v_j}{\partial x_i} - 2(1 - \beta)\frac{\partial v_k}{\partial x_l} n_k n_l n_i n_j = 0 \quad \text{on } \partial B. \quad (4.3.14)$$

For $\beta \neq 0$ it follows that

$$\frac{\partial v_k}{\partial x_l} n_k n_l n_i n_j = 0 \quad \text{on } \partial B$$

and hence that

$$\frac{\partial v_i}{\partial x_j} + \frac{\partial v_j}{\partial x_i} = 0 \quad \text{on } \partial B. \quad (4.3.15)$$

The last result further implies that $\partial v_j/\partial x_j = 0$ on ∂B and since $\partial v_j/\partial x_j$ is harmonic in B, we have $\partial v_j/\partial x_j \equiv 0$ in B. But $V(v_i) = 0$ and therefore $\partial v_i/\partial x_j + \partial v_j/\partial x_i \equiv 0$ in B, from which we conclude that $\sigma_{ij} \equiv 0$ in B. We have thus proved that the stress vanishes identically when either $\alpha > 1$

---

[13] In the present notation, the stress-strain relation (2.1.8) becomes:

$$\sigma_{ij} = \mu\left\{(\alpha - 1)\frac{\partial v_k}{\partial x_k}\delta_{ij} + \frac{\partial v_i}{\partial x_j} + \frac{\partial v_j}{\partial x_i}\right\}.$$

or $\alpha < -1$ in B. The condition $\alpha > 1$ corresponds to

$$0 < \sigma < \tfrac{1}{2}, \qquad\qquad (4.3.16)$$

while $\alpha < -1$ corresponds to

$$\tfrac{1}{2} < \sigma < 1. \qquad\qquad (4.3.17)$$

On recalling Kirchhoff's original result (Theorem 3.1), we see that the theorem is proved for all values of σ in the range (4.3.3) except $\sigma = \tfrac{1}{2}$. This value of σ requires a separate analysis which is now described.

Bramble and Payne [1963a] have shown that under certain smoothness hypotheses the solution u_i of (2.3.2) regarded as a function of σ, converges with its derivatives in compact subdomains of B to a solution of the system

$$\left.\begin{array}{l} \varDelta u_i = \partial p/\partial x_i \\ \partial u_j/\partial x_j = 0, \end{array}\right\} \quad \text{in } B \qquad\qquad \begin{array}{l}(4.3.18)\\(4.3.19)\end{array}$$

and

$$\left(\frac{\partial u_i}{\partial x_j} + \frac{\partial u_j}{\partial x_i}\right) n_j - p\,n_i = f_i \qquad \text{on } \partial B. \qquad\qquad (4.3.20)$$

Here, p is an unknown pressure term to be determined *a posteriori* and f_i is the prescribed surface traction. To prove uniqueness of this problem, it must merely be shown that a vector v_i satisfying (4.3.18), (4.3.19), and (4.3.20) with $f_i = 0$, implies that $\partial v_i/\partial x_j + \partial v_j/\partial x_i \equiv 0$ in B. To this end, consider the identity

$$\int_B v_i [\varDelta v_i - \partial p/\partial x_i]\, dx = \int_{\partial B} v_i \left[\left(\frac{\partial v_i}{\partial x_j} + \frac{\partial v_j}{\partial x_i}\right) n_j - p\,n_i\right] dS$$
$$- \int_B \frac{\partial v_i}{\partial x_j}\left(\frac{\partial v_i}{\partial x_j} + \frac{\partial v_j}{\partial x_i}\right) dx \qquad\qquad (4.3.21)$$

which immediately reduces to

$$\int_B \left(\frac{\partial v_i}{\partial x_j} + \frac{\partial v_j}{\partial x_i}\right)\left(\frac{\partial v_i}{\partial x_j} + \frac{\partial v_j}{\partial x_i}\right) dx = 0, \qquad\qquad (4.3.22)$$

clearly implying the desired result and the theorem is complete.

The same proof obviously holds in the corresponding incompressible displacement and mixed boundary value problems, thus establishing the assertion made on p. 30 during the proof of Theorem 3.3. Note, however, that in the displacement problem the stress is indeterminate to within an arbitrary uniform hydrostatic pressure.

We remark that Theorem 4.3.2 also holds when the region B lies exterior to a star-shaped surface provided v_i tends uniformly to zero as

$r \to \infty$, where r is the distance measured from some origin inside B. We then know from Section 4.2 that

$$v_i = O(r^{-1})$$
$$\partial v_i / \partial x_j = O(r^{-2})$$

(4.3.23)

and so the contribution from spheres of radius r tends to zero as $r \to \infty$. Similarly, we could also require the stress components to vanish uniformly as $r \to \infty$ and still have uniqueness.

By means of arguments similar to those of Theorem 4.3.2, Bramble and Payne [1962b] remove the requirement that B be star-shaped and prove that the traction boundary value problem has at most one solution provided

$$-1 < \sigma < 1 - \tfrac{1}{2} K (1+K)^{-1}, \quad \mu \neq 0,$$

(4.3.24)

where K, a computable positive constant, depends on the geometry of B. Hill [1961a] had earlier given the sufficient condition

$$-1 < \sigma \leqq \tfrac{1}{2} + n, \quad 0 \leqq n < \tfrac{1}{2}, \quad \mu \neq 0,$$

(4.3.25)

for uniqueness in the same problem. Here, n is the lower bound of a certain Rayleigh quotient but it has not yet been exactly determined. The extension of these results to exterior domains follows exactly the same precedure outlined above.

4.3.3 The Traction Boundary Value Problem for a Homogeneous Isotropic Elastic Sphere.
We have already seen in Section 4.1.6 that the displacement boundary value problem for a homogeneous isotropic elastic sphere fails to be unique for certain discrete values of Poisson's ratio. A similar situation occurs in the corresponding traction boundary value problem. The results, due to the Cosserats [1901b, d] and Bramble and Payne [1961a] are that, for domains bounded externally by a spherical surface, the traction boundary value problem fails to be unique[14] whenever the value of Poisson's ratio is 1 or is given by

$$\sigma_n = -[n^2 + n + 1][1 + 2n]^{-1}, \quad n = 0, 1, 2, \dots .$$

(4.3.26)

All the values of σ_n lie in the interval

$$-\infty < \sigma \leqq -1.$$

(4.3.27)

On the other hand, for regions external to a spherical surface, neither the displacement nor the stress is unique for

$$v_i = O(r^{-1}) \quad \text{as } r \to \infty,$$

---

[14] The stress is not unique for values of σ given either by 1 or by (4.3.26) with $n \neq 0$, i.e., in the range $-\infty < \sigma < -1$. Boggio [1907b, p. 448] briefly remarks on the case $\sigma = -1$.

(r being the distance from the origin) provided Poisson's ratio takes on one of the values

$$\sigma_m = [m^2 + 3m + 3][2m + 3]^{-1}, \quad m = 0, 1, 2, \ldots. \quad (4.3.28)$$

The values of σ_m lie in the interval

$$1 \leqq \sigma < \infty. \quad (4.3.29)$$

An obvious conjecture, confirmed in Chapter 6, is that the traction boundary value problem for the homogeneous isotropic elastic half-space has a unique solution unless $\sigma = 1$, this value being the intersection of exceptional values of Poisson's ratio for the interior and exterior sphere problems. The Cosserats [1901d] also established that for a spherical shell there is non-uniqueness at an infinite discrete set of values of Poisson's ratio in the interval

$$-\infty < \sigma \leqq -1, \quad 1 \leqq \sigma < \infty. \quad (4.3.30)$$

Note that (4.3.30) is the sum of the intervals (4.3.27), (4.3.29).

4.3.4 Necessary Conditions for Uniqueness in the Traction Boundary Value Problem for Three-Dimensional Homogeneous Isotropic Elastic Bodies.
Unlike the displacement boundary value problem, a complete set of necessary conditions has not been found for uniqueness in the traction boundary value problem of homogeneous isotropic elastostatics. The results for spherical regions given in the last section together with Theorem 4.3.1 and the theorem of Edelstein and Fosdick on p. 41 suggest that for bounded domains there is uniqueness of stress[15] in the traction problem provided

$$-1 \leqq \sigma < 1. \quad (4.3.31)$$

This conjecture is strengthened by the example of E. and F. Cosserat [1901c] in which there is non-uniqueness for some ellipsoidal region for every value of Poisson's ratio in the range

$$-\infty < \sigma < -1.$$

However, there is no conclusive proof that conditions (4.3.31) are necessary. As might be expected, the position for unbounded domains is even less decisive. For exterior regions, results for a sphere and Theorems 4.3.1 indicate that uniqueness holds if

$$-\infty < \sigma < 1 \quad (4.3.32)$$

and provided either the displacement or stress uniformly approaches its value at infinity.

---

[15] Uniqueness of displacement (up to a rigid body motion) may occur when $-1 < \sigma < 1$.

For regions whose boundaries extend to infinity, Mindlin, referring to work by Neuber, has produced examples in which the traction boundary value problem fails to have a unique solution for values of σ in either of the intervals $-\infty < \sigma \leq -1$ or $1 \leq \sigma < \infty$.

4.4 Mixed Boundary Value Problems

For homogeneous isotropic elastic bodies occupying bounded three-dimensional regions, Kirchhoff's theorem guarantees uniqueness in all the standard boundary value problems including those of mixed type. When the body occupies an exterior region, we have seen that an extension of the theorem is obtained by means of arguments described in Section 4.2. Previous sections have also revealed improvements of the Kirchhoff result when the displacement and traction boundary value problems are separately considered. We now wish to examine more closely certain mixed boundary value problems in order to obtain similar generalisations. While there are several possible types of mixed boundary conditions, we limit consideration to the following four kinds:

(a) The displacement is prescribed on a portion $\overline{\partial B_1}$ of the surface ∂B, and the traction is prescribed on the remainder ∂B_2. Here, $\partial B = \overline{\partial B_1} \cup \partial B_2$, and $\overline{\partial B_1}$ denotes the closure of ∂B_1.

(b) The tangential component of the traction and the normal component of the displacement are prescribed on the entire surface ∂B.

(c) The tangential component of the displacement and the normal component of the traction are prescribed on the entire surface ∂B.

(d) The linear combination $\sigma_{ij} n_j + \beta u_i$ is prescribed on the surface ∂B. Here, β is a non-negative function prescribed on ∂B while the remaining symbols have the usual meaning. This is the so-called "elastic support" condition.

The problems corresponding to these conditions are referred to as problems (a), (b), (c) and (d), respectively.

We first treat the non-homogeneous anisotropic body in a bounded three-dimensional region and obtain a condition on the elasticities sufficient for uniqueness. This reduces to the Kirchhoff criterion in the isotropic case. For problem (a) we show that when supplemented by $\sigma = -1$, this is also necessary for uniqueness in bounded domains. Other generalisations of Kirchhoff's theorem are established for homogeneous isotropic bodies, outside certain convex surfaces and subjected either to conditions (b) or (c). Based on a theorem by Hayes [1964], a number of results are generated for problem (a) in the case of special geometries. These lead to further uniqueness criteria in the displacement and traction boundary value problems for three-dimensional bodies of particular shape.

For simplicity, during the course of this section, only the classical solution is analysed, but the same results are valid for the properly formulated weak solution.

4.4.1 General Anisotropy. The methods used in Sections 4.1 and 4.3 are easily modified to establish the following theorem:

Theorem 4.4.1. *There is at most one solution to the mixed boundary value problem* (a), (b), (c) *or* (d) *for the non-homogeneous anisotropic body occupying a bounded region B, provided the elasticities obey the positive-definiteness condition* (4.3.1). *For spherical domains the solution to problem* (b) *is determined only to within an arbitrary rigid body rotation.*

To prove the theorem we assume, as before, that there exist two solutions u_i^1 and u_i^2, and put

$$v_i = u_i^1 - u_i^2.$$

But Green's identity (*c.p.* 2.2.5) yields

$$\int_B c_{ijkl} \frac{\partial v_i}{\partial x_j} \frac{\partial v_k}{\partial x_l} dx = \int_{\partial B} v_i n_j c_{ijkl} \frac{\partial v_k}{\partial x_l} dS \qquad (4.4.1)$$

which vanishes under conditions (a), (b) and (c) and for (d) reduces to

$$-\beta \int_{\partial B} v_i v_i \, dS. \qquad (4.4.2)$$

Thus, in all cases we conclude that

$$0 \leq c_0 \int_B \left(\frac{\partial v_i}{\partial x_j} + \frac{\partial v_j}{\partial x_i} \right) \left(\frac{\partial v_i}{\partial x_j} + \frac{\partial v_j}{\partial x_i} \right) dx \leq \int_B c_{ijkl} \frac{\partial v_i}{\partial x_j} \frac{\partial v_k}{\partial x_l} dx \leq 0, \qquad (4.4.3)$$

from which it follows that $\partial v_i/\partial x_j + \partial v_j/\partial x_i \equiv 0$ in B. This implies that v_i must be of the form (see Eq. (4.3.2))

$$v_i^* = \alpha_i + e_{ijk} \beta_j x_k. \qquad (4.4.4)$$

The theorem is established once it is proved that v_i^* is identically zero. We examine separately the conditions (a), (b), (c) and (d). Condition (a) implies the vanishing of v_i^* on ∂B_1, which in turn implies the identical vanishing of v_i^* in B. Next, condition (b) enables us to write

$$0 = \int_{\partial B} x_j v_i^* n_i \, dS = \int_B v_j^* \, dx,$$

for v_i^* in the form (4.4.4). With the origin taken at the centroid of B, we conclude that α_i must be zero. Hence,

$$v_i^* = e_{ijk} \beta_j x_k \qquad (4.4.5)$$

so that on each point of the boundary ∂B, we have

$$v_i^* n_i = e_{ijk} \beta_j x_k n_i = \tfrac{1}{2} e_{ijk} \beta_j n_i \frac{\partial r^2}{\partial x_k}, \qquad (4.4.6)$$

where r is the distance from the origin. According to condition (b), the normal component of v_i vanishes on ∂B. In general, from (4.4.6) this is possible only if β_j is zero. However, (4.4.6) also shows that in particular $v_i^* n_i$ can vanish provided r is constant at every point on ∂B, i.e., the boundary is spherical.

In condition (c) we note that $v_i^* n_j - v_j^* n_i$ vanishes on ∂B, and so, by the divergence theorem

$$\int_B \left(\frac{\partial v_i^*}{\partial x_j} - \frac{\partial v_j^*}{\partial x_i} \right) dx = 0, \qquad (4.4.7)$$

and thus we may conclude that β_i is zero. Hence, v_i^* can be at most a rigid body translation α_i. In order to satisfy the vanishing of $v_i^* n_j - v_j^* n_i$ on ∂B we must then require that α_i itself be zero.

Finally, we observe that v_i^* produces zero stress. Therefore, in condition (d), on those portions of the boundary where $\beta \neq 0$, we find that v_i^* is zero. (We are assuming, of course, that β does not vanish identically.) Thus, v_i^* must vanish in B. The proof of the theorem is now complete.

4.4.2 A Homogeneous Isotropic Material. The results of the previous section, upon specialisation to isotropy, show that there is uniqueness in all problems (a)–(d), provided

$$-1 < \sigma < \tfrac{1}{2}, \quad \mu > 0.$$

Several improvements are possible. Upon recalling the analysis given on p. 48 for incompressibility ($\sigma = \tfrac{1}{2}$), we see that uniqueness follows in all problems, except (d), provided

$$-1 < \sigma \leq \tfrac{1}{2}, \quad \mu \neq 0, \qquad (4.4.8)$$

since there is uniqueness also when elasticities are negative definite. In problem (d), however, uniqueness holds provided in (4.4.8) we restrict μ to be positive. In problem (a), the theorem of Edelstein and Fosdick quoted on p. 41 ensures uniqueness provided

$$-1 \leq \sigma \leq \tfrac{1}{2}, \quad \mu \neq 0. \qquad (4.4.9)$$

For boundary conditions (a), it may also be established that for bounded regions, conditions (4.4.9) are necessary for uniqueness. As noted on p. 29 it is enough to demonstrate that there is at least one example for which a violation of (4.4.9) leads to non-uniqueness. Consider a homogeneous isotropic elastic shell of internal and external radii a, b, subjected to a radial symmetric displacement field. As is well known, this displacement field is given by

$$(u_r, u_\theta, u_\varphi) = (c_1 r + c_2 r^{-2}, 0, 0)$$

where c_1, c_2 are arbitrary constants, and (r, θ, φ) denote the spherical polar coordinates. Suppose that the displacement and stress vanish on the inner and outer surfaces respectively. This implies the conditions (we assume for the moment $\mu \neq 0$)

$$c_1 a^3 + c_2 = 0,$$

$$b^3 (3\lambda + 2\mu) c_1 - 4\mu c_2 = 0,$$

which yield non-trivial values of c_1, c_2 if, and only if,

$$\mu(3\lambda + 2\mu) = -4\mu^2 a^3/b^3 < 0.$$

We may thus establish non-uniqueness for all values of the moduli in the ranges

$$-\infty < \sigma < -1, \quad \tfrac{1}{2} < \sigma < +\infty, \quad \mu \neq 0.^{16}$$

The previously introduced counter-examples are easily seen to yield non-uniqueness in the remaining case $\mu = 0$.

Other extensions of the primary result are possible for bodies with particular geometries. We examine several of these, starting with a theorem due to Bramble and Payne [1962a].

Theorem 4.4.2.[17] *Suppose the boundary ∂B of a region B has positive Gaussian curvature. Then in the region B^* exterior to ∂B the problem for a homogeneous isotropic material with mixed boundary condition* (b) *has at most one solution which is twice continuously differentiable in B^* and vanishes uniformly at infinity, provided*

$$-\infty < \sigma < \tfrac{1}{2}, \quad 1 < \sigma < \infty, \quad \mu \neq 0. \tag{4.4.10}$$

Let us assume for the moment that the surface ∂B is smooth so that normal coordinates may be introduced along ∂B. Because of linearity, the proof involves the function v_i obeying

$$\Delta v_i + \alpha \frac{\partial^2 v_j}{\partial x_i \, \partial x_j} = 0, \quad \text{in } B^* \tag{4.4.11}$$

---

[16] Edelstein and Fosdick [1968] consider the same problem but from the different viewpoint of showing that strong ellipticity is not necessary for uniqueness in problem (a).
[17] Guha [1965] arrives at this result using the theory of distributions.

and

$$\left.\begin{array}{r} v_i n_i = 0, \\ \sigma_{ij} n_j n_k - \sigma_{kj} n_j n_i = 0, \end{array}\right\} \quad \text{on } \partial B \qquad (4.4.12)$$

where [18]

$$\sigma_{ij} = \mu(\alpha - 1)\frac{\partial v_k}{\partial x_k}\delta_{ij} + \mu\left(\frac{\partial v_i}{\partial x_j} + \frac{\partial v_j}{\partial x_i}\right), \qquad (4.4.13)$$

and

$$\alpha = (1 - 2\sigma)^{-1}. \qquad (4.4.14)$$

At infinity v_i must vanish uniformly so that (see Section 4.2) provided $\alpha \neq -1$, $\mu \neq 0$, its asymptotic behaviour, together with that of σ_{ij}, is governed by

$$v_i = O(r^{-1}), \qquad \sigma_{ij} = O(r^{-2}) \qquad \text{as } r \to \infty. \qquad (4.4.15)$$

We must prove that v_i vanishes identically in B^*. In virtue of (4.4.11)–(4.4.15) Green's identity leads at once to

$$\int_{B^*} \sigma_{ij}\frac{\partial v_i}{\partial x_j} dx = \mu\int_{B^*}\left\{\frac{\partial v_i}{\partial x_j}\left(\frac{\partial v_i}{\partial x_j} + \frac{\partial v_j}{\partial x_i}\right) + (\alpha - 1)\left(\frac{\partial v_i}{\partial x_i}\frac{\partial v_j}{\partial x_j}\right)\right\} dx = 0 \quad (4.4.16)$$

which, if $\mu \neq 0$, may be rewritten as

$$\begin{aligned} \int_{B^*}\Bigg\{\frac{\partial v_i}{\partial x_j}&\left(\frac{\partial v_i}{\partial x_j} - \frac{\partial v_j}{\partial x_i}\right) + 2\left(\frac{\partial v_i}{\partial x_j}\frac{\partial v_j}{\partial x_i} - \frac{\partial v_j}{\partial x_j}\frac{\partial v_i}{\partial x_i}\right) \\ &+ (\alpha + 1)\frac{\partial v_j}{\partial x_j}\frac{\partial v_i}{\partial x_i}\Bigg\} dx = 0. \end{aligned} \qquad (4.4.17)$$

Let us now consider the second term on the left of (4.4.17). An integration by parts gives

$$\begin{aligned} \int_{B^*}\left[\frac{\partial v_i}{\partial x_j}\frac{\partial v_j}{\partial x_i} - \frac{\partial v_j}{\partial x_j}\frac{\partial v_i}{\partial x_i}\right] dx &= \int_{\partial B} v_i\frac{\partial v_j}{\partial x_i} n_j dS \\ &= \int_{\partial B} v_i\frac{\partial}{\partial x_i}(v_j n_j) dS - \int_{\partial B} v_i v_j\frac{\partial n_j}{\partial x_i} dS, \end{aligned} \qquad (4.4.18)$$

the contribution over the sphere at infinity vanishing because of the second of (4.4.15). The first term on the right of (4.4.18) is clearly zero since $v_j n_j$ vanishes on ∂B. By the assumed positivity of the Gaussian curvature κ (Weatherburn [1930, p. 138]), the second integral of (4.4.18) is non-positive and therefore the left-side of (4.4.18) is non-negative. Note that the normal vector is directed toward the interior of B, i.e., outwards

---

[18] Compare equations (2.1.8).

from B^*. Now, provided $(\alpha+1)>0$, Eq. (4.4.17) and (4.4.18) indicate that,

$$\frac{\partial v_i}{\partial x_i}=\frac{\partial v_i}{\partial x_j}\frac{\partial v_j}{\partial x_i}=0, \tag{4.4.19}$$

from which it immediately follows that v_i is the gradient of some harmonic scalar function. The boundary conditions (4.4.12) then imply that v_i must vanish identically in B^*. Uniqueness is thus established provided ∂B is smooth and

$$\alpha+1>0, \quad \mu\neq0. \tag{4.4.20}$$

But conditions (4.4.20) are equivalent to (4.4.10) and so the theorem is proved. If ∂B is piecewise smooth we can approximate it with smooth surfaces and arrive at the same conclusion.

Another theorem concerned with exterior domains, also due to Bramble and Payne [1962a], is the following:

Theorem 4.4.3. *Suppose the boundary ∂B of a region B has positive average curvature. Then in the region B^* exterior to ∂B, provided*

$$-\infty<\sigma<\tfrac{1}{2}, \quad 1<\sigma<\infty, \quad \mu\neq0,$$

the mixed boundary value problem (c) for a homogeneous isotropic material has at most one solution that is twice continuously differentiable in B^* and vanishes uniformly at infinity.

As in the proof of the previous theorem we assume first that ∂B is smooth. It then suffices to show that the function v_i satisfying (4.4.11) in B^*, (4.4.15) at infinity and

$$\left.\begin{array}{c} v_i n_j - v_j n_i = 0, \\ \sigma_{ij} n_i n_j = 0, \end{array}\right\} \quad \text{on } \partial B \tag{4.4.21}$$

vanishes identically in B^*; here σ_{ij} is given by (4.4.13). Clearly, v_i continues to satisfy (4.4.17) but (4.4.18) is now replaced by

$$\int_{B^*}\left(\frac{\partial v_i}{\partial x_j}\frac{\partial v_j}{\partial x_i}-\frac{\partial v_j}{\partial x_j}\frac{\partial v_i}{\partial x_i}\right)dx=\int_{\partial B} v_i\left(n_j\frac{\partial}{\partial x_i}-n_i\frac{\partial}{\partial x_j}\right)v_j\,dS$$

$$=\int_{\partial B} v_k n_k n_i\left(n_j\frac{\partial}{\partial x_i}-n_i\frac{\partial}{\partial x_j}\right)(v_l n_l n_j)\,dS. \tag{4.4.22}$$

Since $n_i n_j$ is symmetric and $n_j\partial/\partial x_i - n_i\partial/\partial x_j$ antisymmetric in i and j, (4.4.22) reduces to

$$\int_{B^*}\left[\frac{\partial v_i}{\partial x_j}\frac{\partial v_j}{\partial x_i}-\frac{\partial v_j}{\partial x_j}\frac{\partial v_i}{\partial x_i}\right]dx=-\int_{\partial B} J v_j n_j v_i n_i\,dS \tag{4.4.23}$$

in which the identities (Weatherburn [1930])

$$n_j \frac{\partial n_j}{\partial x_i} = 0$$

$$J \equiv \frac{\partial n_j}{\partial x_j} = R_1^{-1} + R_2^{-1}$$

have been used. Here, R_1 and R_2 are the principal radii of curvature at a point on ∂B. The combination of (4.4.19) with (4.4.17) gives

$$\int_{B^*} \left\{ \frac{\partial v_i}{\partial x_j} \left(\frac{\partial v_i}{\partial x_j} - \frac{\partial v_j}{\partial x_i} \right) + (\alpha+1) \frac{\partial v_j}{\partial x_j} \frac{\partial v_i}{\partial x_i} \right\} dx - \int_{\partial B} J v_j n_j v_i n_i \, dS = 0. \quad (4.4.24)$$

We now adopt an argument similar to that used in the previous theorems. Suppose J is nowhere positive on ∂B (i.e., B^* lies external to a boundary of positive average curvature). Then, from (4.4.24), provided $(\alpha+1)>0$, it follows that both the divergence and curl of v_i must vanish identically in B^*. This implies that v_i is the gradient of a harmonic scalar function φ. The first boundary condition (4.4.21) now states that the tangential derivative of φ must vanish on ∂B, and furthermore, on those portions of ∂B where $J \neq 0$, we deduce from (4.4.24) that the normal derivative of φ must vanish. Thus φ is determined by Cauchy data on portions of ∂B. But the harmonic Cauchy problem has a unique solution and so φ is constant and hence v_i is identically zero in B^*. This completes the proof of the present theorem for smooth ∂B. Again, a non-smooth boundary may be approximated by smooth ones and the same result obtained.

When the domain B is polygonal, Bramble and Payne [1962a] have further proved that conditions (c) lead to a unique solution in B^* (with suitable restrictions at infinity) provided merely that

$$\sigma \neq 1, \quad \mu \neq 0. \tag{4.4.25}$$

Counter-examples for the exceptional values of σ and μ may be easily given, showing that (4.4.25) in this case are both necessary and sufficient for uniqueness.

Interesting results evolve from a uniqueness theorem constructed by Hayes [1964]. This theorem concerns boundary conditions (a) applied to an anisotropic body occupying a finite region. The portion of the boundary ∂B_1 is star-shaped with respect to an origin located on ∂B_2, i.e.,

$$n_k x_k > 0 \quad \text{on } \partial B_1,$$

while the portion ∂B_2 is planar, i.e.,

$$n_k x_k = 0 \quad \text{on } \partial B_2.$$

Unless otherwise stated, in the remainder of this section ∂B_1 and ∂B_2 have the same meaning as in condition (a).

A slight generalisation of Hayes' theorem is

Theorem 4.4.4. *For the above-mentioned domain, there is at most one solution of the anisotropic equations for the boundary conditions* (a) *provided the elasticities at every point of* $B \cup \partial B_1$ *are analytic, satisfy*

$$b_{ijkl}\, \xi_{ij}\, \xi_{kl} \equiv \left[x_p \frac{\partial c_{ijkl}}{\partial x_p} + \beta\, c_{ijkl} \right] \xi_{ij}\, \xi_{kl} \geqq 0 \qquad (4.1.13)$$

for some constant β *and every tensor* ξ_{ij}, *obey the symmetry* (4.1.11) *and satisfy the strong ellipticity condition*

$$c_{ijkl}\, \alpha_i\, \alpha_k\, \beta_j\, \beta_l \geqq c_1\, \alpha_i\, \alpha_i\, \beta_j\, \beta_j \qquad (4.1.7)$$

for some positive constant c_1 *and every pair of vectors* α_i, β_i.

The proof utilises the technique due to Bramble and Payne [1962a; see Theorem 4.3.2]. Let v_i be the difference of two solutions to the problem. Green's identity then yields

$$\int_B c_{ijkl} \frac{\partial v_i}{\partial x_j} \frac{\partial v_k}{\partial x_l}\, dx = 0. \qquad (4.4.26)$$

On the other hand, from the equilibrium equations (2.1.23), an integration by parts, and the symmetry condition (4.1.11), we get

$$0 = \int_B x_p \frac{\partial v_i}{\partial x_p} \frac{\partial}{\partial x_j} \left(c_{ijkl} \frac{\partial v_k}{\partial x_l} \right) dx$$

$$= \int_{\partial B} x_p \frac{\partial v_i}{\partial x_p} \sigma_{ij}\, n_j\, dS - \frac{1}{2} \int_{\partial B} x_p\, n_p\, c_{ijkl} \frac{\partial v_i}{\partial x_j} \frac{\partial v_k}{\partial x_l}\, dS \qquad (4.4.27)$$

$$+ \frac{1}{2} \int_B b_{ijkl} \frac{\partial v_i}{\partial x_j} \frac{\partial v_k}{\partial x_l}\, dx.$$

The boundary integrals clearly vanish on ∂B_2, since $\sigma_{ij}\, n_j$ and $x_k\, n_k$ are zero there. On ∂B_1, v_i vanishes, and consequently (4.4.27) reduces to

$$\int_{\partial B_1} x_p\, n_p\, c_{ijkl} \frac{\partial v_i}{\partial n} \frac{\partial v_k}{\partial n}\, n_j\, n_l\, dS \leqq 0 \qquad (4.4.28)$$

where $\partial v_i / \partial n$ denotes the normal derivative of v_i on ∂B_1. By hypothesis, the elasticities satisfy the strong ellipticity condition (4.1.7) at each point of ∂B_1, implying from (4.4.28) that not only v_i but $\partial v_i / \partial n$ must vanish on ∂B_1. But the elasticities are also analytic in $B \cup \partial B_1$ and therefore an appeal to Holmgren's uniqueness theorem completes the proof.

Observe that the elasticities need not be analytic on the whole of ∂B_1. It is sufficient that they be analytic in B and extendable as analytic functions in the neighbourhood of some set of non-zero measure on ∂B_1. When the body is homogeneous we may take $\beta = 0$ in (4.1.13).

Hayes [1964] considered only constant elasticities, noting that for isotropy Theorem 4.4.4 gives uniqueness provided

$$-\infty < \sigma < \tfrac{1}{2}, \quad 1 < \sigma < \infty, \quad \mu > 0. \tag{4.4.29}$$

The result is also suggested by the reflexion principles of Duffin [1956]. These principles immediately indicate an analogous theorem in which ∂B_1 and ∂B_2 are interchanged, a result easily verified using the procedure of proof in the previous theorem.

Theorem 4.4.5. *For a bounded region B whose boundary consists of a planar part ∂B_1 with the remainder ∂B_2 being star-shaped with respect to an origin on ∂B_1, there is at most one solution to the homogeneous isotropic problem with boundary conditions (a) provided*

$$-1 < \sigma < 1, \quad \mu \neq 0. \tag{4.4.30}$$

The incompressible case, $\sigma = \tfrac{1}{2}$, must be treated separately, but uniqueness can be easily confirmed.

Note that by means of a counter-example Hayes [1964] has proved that Theorem 4.4.5 cannot be extended to the anisotropic case, at least in the theory of small deformations superposed upon large.

Duffin's reflexion principles [1956] further indicate that the theorem of Hayes should hold for any of the standard boundary conditions prescribed on the plane ∂B_2. The correctness of this assertion can be seen from the identity (4.4.27) where the second boundary integral now vanishes on ∂B_2 (as there $x_k n_k = 0$), while the integrand of the first boundary integral is the product of a tangential derivative of v_i and the i-th component of the surface traction. Thus, all integrals over ∂B_2 vanish whenever v_i satisfies the displacement, traction or mixed boundary conditions of all types except (d). When mixed boundary conditions of type (d) are specified on ∂B_2, the appropriate integral over ∂B_2 reduces to

$$\int_{\partial B_2} x_p \frac{\partial v_i}{\partial x_p} \sigma_{ij} n_j \, dS = -\int_{\partial B_2} \alpha \, x_p \frac{\partial v_i}{\partial x_p} v_i \, dS$$

$$= \frac{1}{2} \int_{\partial B_2} v_i \, v_i \frac{\partial}{\partial x_p} (\alpha \, x_p) \, dS. \tag{4.4.31}$$

Hence, on supposing $x_p \, \partial \alpha / \partial x_p + 2\alpha \geq 0$ on ∂B_2, we may in all cases proceed as in Theorem 4.4.4 which therefore remains valid in the present set of circumstances.

As might be expected, the subsequent theorem in which ∂B_1 and ∂B_2 are interchanged also remains valid. Here, the standard boundary conditions are given on the plane boundary ∂B_1, and the traction on the star-shaped portion ∂B_2.

Other interesting results follow from successive applications of the identity (4.4.27). For instance, suppose an isotropic homogeneous elastic material is bounded by the planes $z = 0$ and $z = l$ and that the displacement vanishes appropriately at infinity. Let the displacement be prescribed on the plane $z = l$ and the tractions on the plane $z = 0$. Hayes' arguments then prove uniqueness, provided $\mu \neq 0$, $-\infty < \sigma < \frac{1}{2}$, $1 < \sigma < \infty$. On the other hand, by choosing the origin on the plane where the displacement is specified, we see that in (4.4.27) the integral over this plane vanishes since $x_p \, \partial v_i / \partial x_p$ is merely the tangential derivative of v_i and $x_k n_k$ is zero. The arguments of Bramble and Payne [see Theorem 4.3.2] may now be applied to the remaining integral in (4.4.27) to establish uniqueness provided $\mu \neq 0$, $-1 < \sigma < +1$. Combining both results, we determine that the mixed boundary value problem in question has a unique solution, provided

$$\sigma \neq 1, \quad \mu \neq 0. \tag{4.4.32}$$

Uniqueness also follows if (4.4.32) is satisfied and the finite region B is bounded by the intersection of a right circular conical surface and a plane surface. The material is assumed isotropic, the displacements are prescribed on the conical surface and the tractions on the plane surface (or vice versa). The conclusion follows as before by taking the origin successively at the vertex of the cone and at a suitable point in the plane face. This result is equally valid when the cone is replaced by a right pyramid. Suppose now the region B is reflected in the boundary plane to form a region B^*. Let the tractions be prescribed on the reflected surface, ∂B^*, with the displacement being prescribed as before. Choosing the origin alternatively at both vertices of $B \cup B^*$ and repeating the previous arguments we see that uniqueness holds again under condition (4.4.32). Finally, consider the finite region bounded by a right circular cone and a surface star-shaped with respect to the origin located at the vertex of the cone. Assume that the tractions are prescribed on the cone and that the displacement is prescribed on the remainder of the surface. The argument of Hayes establishes uniqueness under conditions (4.1.7) or (4.4.29). Assume next that the displacement is prescribed on the conical portion of the surface with the traction specified on the remainder. The argument of Bramble and Payne establishes uniqueness under condition (4.4.30). Of course, we may again replace the cone by a right pyramid and recover uniqueness under the same conditions.

Chapter 5

Uniqueness Theorems in Homogeneous Isotropic Two-Dimensional Elastostatics

The development of uniqueness theorems in homogeneous isotropic two-dimensional elastostatics, although begun much later than in the three-dimensional theory, nevertheless rapidly achieved a state of greater completion. Kirchhoff's theorem has an obvious counterpart in the plane theory, but by 1933 Muskhelisvili had considerably advanced this result in both the displacement and traction boundary value problems. An important paper published in that year contains a section in which he derived sufficient conditions for uniqueness in the displacement boundary value problem analogous to those previously obtained by the Cosserats in the three-dimensional theory. In addition he proved that for the traction boundary value problem the stress field is unique provided only that $\sigma \neq 1$ and $\mu \neq 0$. In the same paper Muskhelisvili also succeeded in establishing reasonable conditions on the asymptotic behaviour of the displacement and stress fields at infinity which are sufficient to insure that uniqueness continues to hold in problems for exterior domains. Moreover, the techniques he later announced for solving particular problems produce complete solutions, and therefore yield corresponding uniqueness criteria. This further widens the knowledge in the plane theory.

The present chapter contains an account of these results. We give first the modified form of Kirchhoff's theorem with its extension to exterior domains, and then establish necessary and sufficient conditions for uniqueness in the displacement and traction boundary value problems. We end by considering bodies with special geometries, dealing explicitly with interior and exterior regions bounded by a circle, but leaving until the next chapter problems concerned with the whole and half-plane. Further results can be obtained from appropriate sections of Muskhelisvili's book.

Throughout this chapter only classical solutions are considered. In the plane anisotropic theory, results strictly analogous to those in three dimensions are possible, and for this reason their discussion is omitted.

In addition to the unbounded two-dimensional regions treated in the present text, Tiffen [1952] has considered others that include the infinite slab and the region whose boundary (not necessarily planar) extends to infinity. In addition to conditions on the moduli, he stipulates what asymptotic behaviour at infinity the displacement and stress fields must obey in order that the various uniqueness theorems remain valid.

An appendix to this chapter deals briefly with the uniqueness of the solution in the closely related three-dimensional axisymmetric problem. Conditions for uniqueness are actually identical to those in the corresponding isotropic plane boundary value problem, and their derivation is described without proof.

5.1 Kirchhoff's Theorem in Two-Dimensions. The Displacement and Traction Boundary Value Problems

We use the plane strain theory governed by Eq. (2.3.10) and prove the following theorem.

Theorem 5.1.1 (Kirchhoff). *There is at most one solution to the isotropic standard boundary value problems of plane strain provided*

$$-\infty < \sigma < \tfrac{1}{2}, \quad \mu \neq 0. \tag{5.1.1}$$

In the traction boundary value problem there is uniqueness to within a rigid body displacement.

The proof follows traditional lines. Let u_i^1 and u_i^2 be two solutions to the problem and set

$$v_i = u_i^1 - u_i^2.$$

Hence, v_i satisfies the equations of the homogeneous problem and Green's identity accordingly yields,

$$\int_D \left[\lambda \frac{\partial v_\alpha}{\partial x_\alpha} \frac{\partial v_\beta}{\partial x_\beta} + \frac{\mu}{2} \left(\frac{\partial v_\alpha}{\partial x_\beta} + \frac{\partial v_\beta}{\partial x_\alpha} \right) \left(\frac{\partial v_\alpha}{\partial x_\beta} + \frac{\partial v_\beta}{\partial x_\alpha} \right) \right] dx = 0 \tag{5.1.2}$$

where α, β have the range 1, 2, and D is the area of the cross-section of the body. The integrand in (5.1.2) is sign-definite if (and only if) (5.1.1) holds. Then (5.1.2) yields

$$\frac{\partial v_\alpha}{\partial x_\beta} + \frac{\partial v_\beta}{\partial x_\alpha} = 0$$

from which the conclusion follows.

Uniqueness also holds in the incompressible case, but must be established directly from the modified equations as indicated in Section 4.2.2. Moreover, an extension of Theorem 5.1.1 and others of this chapter to exterior domains follows immediately from Muskhelisvili's work (see [1953] § 40), subject to the requirement that either the displacement or stress must approach uniformly its prescribed value at infinity. In the case of prescribed stress, we must give in addition the resultant force over all internal boundaries together with the rotation at infinity.

For the displacement boundary value problem it is possible to establish the following theorem, which is the analogue of Theorem 3.3.

Theorem 5.1.2. *There is at most one solution to the interior displacement boundary value problem in the homogeneous isotropic elastic theory of plane strain if and only if*

$$-\infty \leqq \sigma \leqq \tfrac{1}{2}, \quad 1 < \sigma \leqq \infty, \quad \mu \neq 0. \tag{5.1.3}$$

This theorem is essentially due to Hill [1961a], although the sufficiency part had earlier been proved by Muskhelisvili [1933]. Indeed, the proof of sufficiency is entirely analogous to the three-dimensional procedure described in Theorem 3.3, and so need not be repeated. Necessity is demonstrated by means of a counter-example. Consider the displacement field given by (cf. Cosserats [1898c], Ericksen [1957])

$$u_\alpha = (a^2 x^2 + b^2 y^2 - 1)\,\delta_{1\alpha}, \quad \alpha = 1, 2,$$

which vanishes on the ellipse $a^2 x^2 + b^2 y^2 = 1$ and satisfies Navier's equations provided

$$(\lambda + 2\mu)\,\mu = -\mu^2\,b^2/a^2 < 0.$$

Suitable variation of a^2 and b^2 then shows existence of non-unique solutions for all values of σ not included in the range (5.1.3). Again, when $\mu = 0$ it may be shown, as before, that any solenoidal displacement field vanishing on the boundary is a solution, and therefore non-uniqueness follows trivially. The theorem is thus proved.

Extension of Theorem 5.1.2 to exterior domains follows under conditions outlined after the proof of Theorem 5.1.1. It should also be observed that for $\sigma = \tfrac{1}{2}$ and $\sigma = \pm\infty$ uniqueness continues to hold in conformity with Theorem 3.3.

Much more can be proved in the traction boundary value problem, as the following theorem shows.

Theorem 5.1.3. *For a homogeneous isotropic material occupying a bounded region D, the necessary and sufficient condition for a unique stress distribution in the traction boundary value problem of plane strain is*

$$\sigma \neq 1. \tag{5.1.4}$$

The displacement, assumed single-valued in D, is uniquely defined to within a rigid body displacement if, and only if,

$$\sigma \neq 1, \quad \mu \neq 0. \tag{5.1.5}$$

We suppose that the region D is multiply-connected, bounded externally by the simple closed curve ∂D_0, and internally by the simple closed curves ∂D_m $(m = 1, \ldots, n)$. Necessity is trivially established. For $\sigma = 1$, $\mu \neq 0$, a non-unique solution is given by[1]

$$v_\alpha = \partial \varphi / \partial x_\alpha, \quad \alpha = 1, 2,$$

where $\partial \varphi / \partial x_\alpha$ is chosen to vanish on $\partial D = \bigcup_{m=0}^{n} \partial D_m$. For $\sigma \neq 1$, $\mu = 0$, the stress vanishes identically (hence is unique) while any solenoidal displacement vector is a solution. The necessity of (5.1.4) and (5.1.5) for uniqueness of stress and displacement is therefore proved. In discussing sufficiency, we may accordingly assume without loss that $\mu \neq 0$.

The sufficient condition was first discovered by Muskhelisvili [1933] utilising a method based upon the Airy stress function and complex analysis.[2] A more direct proof, later supplied by Hill [1961a], is the basis for the following. The strain energy may be written

$$V = \frac{1}{4\mu} \int_D \{(1-\sigma)(\sigma_{11}+\sigma_{22})^2 + 2(\sigma_{12}^2 - \sigma_{11}\sigma_{22})\} \, dx, \tag{5.1.6}$$

and the theorem is established once the second term on the right is shown to be zero. Now, from a well-known theorem in calculus, it follows that for multiply connected regions the necessary and sufficient conditions for the equilibrium equations to hold in the absence of body force is that there exist functions $G_1(x_1, x_2)$ and $G_2(x_1, x_2)$, possibly multiple valued, which are related to the stresses by

$$\sigma_{11} = \frac{\partial G_1}{\partial x_2}, \quad \sigma_{22} = \frac{\partial G_2}{\partial x_1}, \quad \sigma_{12} = -\frac{\partial G_1}{\partial x_1} = -\frac{\partial G_2}{\partial x_2}.$$

Consider any curve C connecting the point (x_1, x_2) to an arbitrary origin $(\bar{x}_1, \bar{x}_2)$ and lying entirely in D. Then by direct substitution for the stresses it may be shown that

$$\int_C (\sigma_{\alpha 1} n_1 + \sigma_{\alpha 2} n_2) \, ds = G_\alpha(x_1, x_2) - G_\alpha(\bar{x}_1, \bar{x}_2), \quad \alpha = 1, 2, \tag{5.1.7}$$

where s is the arc-length along the curve. Hence, the functions G_α may be interpreted as the cartesian components of the resultant traction across

---

[1] Compare p. 28 and the references mentioned there.
[2] See also Sherman [1938] who derived the same result from an integral equation formulation.

the curve C. We next consider the second term on the right of (5.1.6). We have, on using Green's theorem,

$$
\begin{aligned}
\int_D (\sigma_{11}\sigma_{22}-\sigma_{12}^2)\,dx &= \int_D \left[\frac{\partial}{\partial x_1}\left(G_1\frac{\partial G_2}{\partial x_2}\right)-\frac{\partial}{\partial x_2}\left(G_1\frac{\partial G_2}{\partial x_1}\right)\right]dx \\
&= \oint_{\partial D} G_1\,dG_2 = -\oint_{\partial D} G_2\,dG_1.
\end{aligned}
\tag{5.1.8}
$$

For the purpose of proving uniqueness, the traction may be assumed zero on each contour ∂D_m. From (5.1.7) it then follows that G_α is constant on each ∂D_m and therefore single valued in D. Hence, by (5.1.8), we get

$$
\int_D (\sigma_{11}\sigma_{22}-\sigma_{12}^2)\,dx=0.
\tag{5.1.9}
$$

It now follows immediately that

$$
\int_D [(\sigma_{11}-\sigma_{22})^2+4\sigma_{12}^2]\,dx=\int_D (\sigma_{11}+\sigma_{22})^2\,dx.
$$

Thus with the help of (5.1.9) the strain energy may be written as

$$
V\equiv\frac{1}{4\mu}\int_D (1-\sigma)[(\sigma_{11}-\sigma_{22})^2+4\sigma_{12}^2]\,dx.
\tag{5.1.10}
$$

A standard application of Green's theorem also shows that the strain energy given by (5.1.6) is identically zero. Consequently by (5.1.6) and (5.1.10) all stresses vanish provided $\sigma\neq1$. The theorem is thus proved.

An extension of Theorem 5.1.2 to exterior domains may be easily accomplished using the results of Muskhelisvili [1953, § 40]. Uniqueness is guaranteed in this instance under the same conditions as in the theorem provided that, in addition, either the displacement or stress approaches uniformly its prescribed value at infinity. For the case of prescribed stress at infinity, we must, in addition, prescribe the rotation there. Uniqueness in the plane traction boundary value problem is therefore completely resolved.

5.2 Uniqueness in Plane Problems with Special Geometries

We start with some comments on bodies possessing special geometries, and then give a complete discussion of uniqueness in the displacement boundary value problem for the region bounded by a circle. Before proceeding to this proof, however, let us first note that Muskhelisvili, in his book [1953], has described techniques for solving standard boundary value problems in regions with plane, circular and elliptic boundaries. Since his methods lead to *complete* solutions of the respective problems, these may be used in proving a corresponding uniqueness theorem in the following manner: take the complete solution and from it determine for what values of the elasticities it becomes identically zero under homogeneous data. This provides sufficient conditions for uniqueness. Counter-examples, usually suggested by the method of solution itself, can be easily constructed for the exceptional values in order to obtain necessary conditions. In this way, it may be proved that necessary and sufficient conditions for a unique solution in, for instance, the exterior displacement boundary value problem for an ellipse, are $\sigma \neq \frac{3}{4}$, 1, $\mu \neq 0$, provided the displacement vanishes uniformly at infinity.[3] The value $\sigma = 1$ is excluded because at this value the entire complex variable formulation of plane elasticity fails and therefore also the method of solution adopted by Muskhelisvili. However, in this case, non-uniqueness may be demonstrated by means of a counter-example similar to that introduced in the proof of Theorem 5.1.2.

We now turn our attention to another method of proving uniqueness in regions of special geometry, and illustrate the approach by considering explicitly the displacement boundary value problems for plane regions interior and exterior to a circle. A similar treatment may be applied to problems in the whole and half plane, and an example is given in the next chapter. The method makes no use of complex variables.

Thus, for the region bounded by a circle, we see that when $\sigma \neq 1$, the displacement is biharmonic and therefore may be represented by

$$u_\alpha = h_\alpha + (r^2 - a^2)\, H_\alpha, \qquad r^2 = x_\alpha x_\alpha, \tag{5.2.1}$$

where h_α, H_α ($\alpha = 1, 2$) are harmonic functions in $r \leq a$, and a is the radius of the disc whose centre is located at the origin. To prove uniqueness, we assume as before that two displacement vectors exist and take their difference, v_i. Then in both the interior and exterior problems (on assuming that $\sigma \neq 1$, $\mu \neq 0$), we see that v_i is of the form (5.2.1). Also v_i vanishes on $r = a$. We consider first the interior problem. It follows immediately that $h_\alpha{}^4$ vanishes identically in $r \leq a$, whereupon Eq. (5.2.1) yields the expressions

$$\partial v_\alpha / \partial x_\alpha = 2 x_\beta H_\beta + (r^2 - a^2)\, \partial H_\alpha / \partial x_\alpha$$
$$\partial v_\alpha / \partial x_\beta - \partial v_\beta / \partial x_\alpha = 2(x_\beta H_\alpha - x_\alpha H_\beta) + (r^2 - a^2)(\partial H_\alpha / \partial x_\beta - \partial H_\beta / \partial x_\alpha). \tag{5.2.2}$$

---

[3] See Muskhelisvili [1953, § 83].
[4] See Bramble [1960] and Duffin [1955].

Since both quantities on the left of (5.2.2) are harmonic, we further get

$$2\,\partial H_\alpha/\partial x_\alpha + x_\beta\,\partial^2\,H_\alpha/\partial x_\alpha\,\partial x_\beta = 0,$$
$$2(\partial H_\alpha/\partial x_\beta - \partial H_\beta/\partial x_\alpha) + x_\gamma\,\partial(\partial H_\alpha/\partial x_\beta - \partial H_\beta/\partial x_\alpha)/\partial x_\gamma = 0. \tag{5.2.3}$$

The solution of each of these equations is either zero or of the form $g(\theta)\,r^{-\frac{1}{2}}$. The latter is impossible since H is harmonic in $r \leq a$. Thus

$$\partial H_\alpha/\partial x_\alpha = \partial H_\alpha/\partial x_\beta - \partial H_\beta/\partial x_\alpha = 0$$

and therefore

$$H_\alpha = \partial H/\partial x_\alpha, \tag{5.2.4}$$

for some harmonic function H. Substitution of the modified expression for the displacement into Navier's equation leads to

$$(2+\kappa)\,\partial(x_\beta\,\partial H/\partial x_\beta)/\partial x_\gamma = 0, \qquad \kappa = 1/(1-2\sigma).$$

An integration gives

$$(2+\kappa)\,H = k\,\log r + f(\theta), \tag{5.2.5}$$

where f is a harmonic function of θ and k is some constant. Obviously, since u is continuous in $r \leq a$, $f = k = 0$.[5] Thus, from (5.2.5), provided $2+\kappa \neq 0$ (i.e., $\sigma \neq \frac{3}{4}$), we may conclude that $H \equiv 0$ and therefore uniqueness follows. If $\sigma = \frac{3}{4}$, $\mu \neq 0$, a simple counter-example demonstrating non-uniqueness is

$$2\mu\,v_\alpha = (x_\beta\,x_\beta - 1)\,\delta_{1\alpha}.$$

Counter-examples for $\sigma = 1$, $\mu \neq 0$, and for $\sigma \neq 1$, $\frac{3}{4}$, $\mu = 0$, may be constructed as before, thus showing that *necessary and sufficient conditions for uniqueness in the plane isotropic interior displacement boundary value problem for a circular disc are $\sigma \neq 1$, $\frac{3}{4}$, $\mu \neq 0$.*

In the exterior problem, we assume that at infinity the displacement and stress are subject to either of the conditions used in extending the plane form of Kirchhoff's theorem. Both imply the asymptotic behaviour

$$v_\alpha = O(r^{-1}), \qquad \partial v_\alpha/\partial x_\beta = O(r^{-2}) \qquad \text{as } r \to \infty. \tag{5.2.6}$$

The proof of uniqueness differs from that in the interior problem since h_α cannot be proved zero at the outset. However, the retention of h_α still leads in the same manner to Eqs. (5.2.3), and hence to (5.2.4). On replacing H_α by $\partial H/\partial x_\alpha$ in (5.2.1) and substituting in Navier's equation, we get, after an integration, the equation

$$2(2+\kappa)\,x_\beta\,\partial H/\partial x_\beta + \kappa\,\partial h_\beta/\partial x_\beta = k \tag{5.2.7}$$

---

[5] The additive constant in the expression for $f(\theta)$ may be taken to be zero without loss.

for constant k. Define harmonic functions $\hat{h}_\alpha$ and $\hat{H}$ by

$$\hat{h}_\alpha = h_\alpha + k\, x_\alpha/4, \qquad \hat{H} = H - (k\,\log r)/4 \qquad (5.2.8)$$

so that (5.2.7) becomes

$$2(2+\kappa)\,x_\beta\,\partial\hat{H}/\partial x_\beta + \kappa\,\partial\hat{h}_\beta/\partial x_\beta = 0. \qquad (5.2.9)$$

Assume, for the moment, that $\kappa \neq 0$. Then using (5.2.9) we see that the dilatation is given by

$$\partial v_\beta/\partial x_\beta = 2x_\beta\,\partial\hat{H}/\partial x_\beta + \partial\hat{h}_\beta/\partial x_\beta = -\frac{4}{\kappa}\,x_\beta\,\partial\hat{H}/\partial x_\beta. \qquad (5.2.10)$$

Now, because $\hat{H}$ is harmonic it is of the same order at infinity as $x_\beta\,\partial\hat{H}/\partial x_\beta$, and therefore (5.2.10) with (5.2.6) show that

$$\partial\hat{H}/\partial x_\alpha = O(r^{-3}) \qquad \text{as } r \to \infty. \qquad (5.2.11)$$

Let us now use (5.2.8) to introduce $\hat{h}_\alpha$ and $\hat{H}$ in the representation (5.2.1) for v_α. We get

$$v_\alpha = \hat{h}_\alpha - \frac{k\,a^2\,x_\alpha}{4r^2} + (r^2 - a^2)\,\partial\hat{H}/\partial x_\alpha, \qquad (5.2.12)$$

which shows that the harmonic function, $(\hat{h}_\alpha - k\,a^2\,x_\alpha/4r^2)$, vanishes on $r = a$. But by (5.2.6) and (5.2.11) this function is $O(r^{-1})$ as $r \to \infty$ and is therefore identically zero. Thus, we have

$$\hat{h}_\alpha = \frac{k\,a^2\,x_\alpha}{4r^2}, \qquad r \geq a.$$

A substitution for $\hat{h}_\alpha$ in (5.2.9) followed by an integration next gives

$$2(2+\kappa)\,\hat{H} = 0,$$

where the function of integration, dependent upon the polar angle alone, is zero because of the behaviour of $\hat{H}$ at infinity. Thus, if $2+\kappa \neq 0$, we conclude uniqueness.

We now consider the case $\kappa = 0$. From the Navier equations it follows that v_α is harmonic. Condition (5.2.6) then implies that v_i vanish identically and uniqueness again follows. If $2+\kappa = 0$ (i.e., $\sigma = \frac{3}{4}$) a counter-example to uniqueness is

$$2\mu v_1 = \frac{(x^2+y^2-a^2)}{(x^2+y^2)^3}[x^3 - 3x\,y^2],$$

$$2\mu v_2 = \frac{(x^2+y^2-a^2)}{(x^2+y^2)^3}[-3x^2\,y + y^3].$$

Counter-examples for the other exceptional values ($\sigma = 1$, $\sigma = \frac{3}{4}$ or $\mu = 0$) of the elasticities can be quite easily constructed. Thus, we have proved that the *homogeneous isotropic plane displacement boundary value problem in a region exterior to a circle has at most one solution if and only if $\sigma \neq 1, \frac{3}{4}, \mu \neq 0$, and the displacement approaches uniformly its prescribed value at infinity. Alternatively, the stress may approach its prescribed value at infinity, provided the rotation vanishes there, and the resultant force over the inner circle is prescribed.*

Criteria in both the interior and exterior problems are considerably simpler than their three-dimensional analogues where, as described in Section 4.1.6, uniqueness fails at an infinite number of discrete values of Poisson's ratio.

Appendix
Uniqueness of Three-Dimensional Axisymmetric Solutions

We state briefly, without proof, conditions for the uniqueness of the classical solution to three-dimensional axisymmetric problems in homogeneous isotropic elastostatics. A method of obtaining these criteria depends upon a transformation, due originally to Weber [1925, 1940] and later developed by Aleksandrov [1959, 1961], that converts the displacement and stress of a plane strain solution into those of a related axisymmetric solution. Knops [1965a] proved that the axisymmetric solution generated in this fashion is complete and bears a one-one correspondence to the plane solution. By this means he transferred sufficient conditions for uniqueness from the plane to the axisymmetric problem. Necessary conditions were obtained from counterexamples. He found that necessary and sufficient conditions for uniqueness in the axisymmetric displacement boundary value problem are given by (5.1.3), while in the axisymmetric traction boundary value problem they are given by (5.1.4) and (5.1.5). Knops [1966] treated the exterior problem by the same technique and derived similar results.

Chapter 6
Problems in the Whole- and Half-Space

Among comparatively recent contributions to elastostatic uniqueness theorems are those concerned with problems of the whole- and half-space. Their late arrival may appear surprising in view of the practical importance of such problems and the length of time they have been studied, yet because in the half-space the extension of Kirchhoff's classical theorem is more difficult than in exterior domains, the appropriate uniqueness theorems have accordingly taken longer to establish. On the other hand, the situation in the whole space is almost trivial and probably for this reason it has never been recorded. Of course, it is easy to state the requisite order conditions that the displacement and stress must satisfy at infinity in order to make the pertinent integrals converge to zero in the application of Kirchhoff's theorem, but as outlined in Section 4.2, these prescriptions are at best artifical. The difficulty arises in determining the rate of growth or decay implied by suitably mild restrictions on the asymptotic behaviour of the displacement and stress components. Actually, we shall show that once this knowledge is obtained, the classical energy arguments become unnecessary, and instead we may use, for instance, Duffin's reflexion principles to derive uniqueness.

The *reflexion principles* established by *Duffin* [1956] concern a homogeneous, isotropic, elastic material, occupying the half-space $x_1 > 0$, and demonstrate how the solution of the equilibrium equations may be analytically continued across the plane $x_1 = 0$ under a variety of homogeneous boundary conditions at $x_1 = 0$. Explicitly, they state the following: let P^* be an open set such that if (x_1, x_2, x_3) is a point of P^* so also is $(-x_1, x_2, x_3)$. Let Q be the intersection of P^* with the plane $x_1 = 0$, and let P be the intersection of P^* with the half-space $x_1 > 0$. Suppose that u_i $(i = 1, 2, 3)$ is the displacement field satisfying the Navier equations (2.3.2) in P. Then if $\mu \neq 0$ these functions have a continuation which satisfies (2.3.2) in all of P^* provided that either

(a) u_i vanish as any point of Q is approached and $\sigma \neq \frac{3}{4}, 1$,

or (b) the stresses $\sigma_{11}, \sigma_{12}, \sigma_{13}$ vanish as any point of Q is approached, $\sigma \neq 1$, and the line segment joining (x_1, x_2, x_3) to $(-x_1, x_2, x_3)$ lies entirely in P^*,

or (c) the displacement component u_1 and the stress components σ_{12}, σ_{13} vanish as any point of Q is approached and $\sigma \neq 1$,

or (d) the displacement components u_2, u_3 and the stress component σ_{11} vanish as any point of Q is approached and $\sigma \neq 1$.

Proofs of all the above statements are given by Duffin [1956]. For other continuation formulae in elasticity see Bramble [1960, 1961] and Bramble and Payne [1961 b, 1961 c].

Apart from Kirchhoff's result which involved restrictive hypotheses, little was known about uniqueness in half-space problems until the appearance of a book by Mikhlin [1962, p. 220] on multidimensional singular integral equations. After proving that a solution to the general homogeneous isotropic displacement and traction boundary value problems is usually possible, except when Poisson's ratio equals $\frac{1}{2}$, $\frac{3}{4}$, 1, Mikhlin constructed a counter-example demonstrating the lack of uniqueness in the half-space displacement problem when Poisson's ratio is $\frac{3}{4}$. Sufficient conditions for uniqueness were not derived. The next result, given by Knops [1965b], also concerned the displacement problem and established the fact that

$$\sigma \neq 1, \quad \sigma \neq \tfrac{3}{4}; \quad \mu \neq 0 \tag{6.1}$$

are necessary and sufficient for uniqueness. This proof was not completely satisfactory since unnecessary assumptions were imposed on the continuity of the dilatation onto the boundary and on the behaviour at infinity, while the displacement was required to satisfy smoothness conditions beyond those ordinarily needed. These extraneous assumptions were removed in a subsequent proof by Turteltaub and Sternberg [1967] who also succeeded in extending their treatment to include the traction problem in the half-space. These authors require that in the displacement problem any two displacement fields approach the same continuous surface values and their difference remains bounded at infinity. In the corresponding traction problem, any two stress fields are required to approach the same continuous surface traction while their difference is required to tend to zero at infinity. Under these respective hypotheses it is confirmed that (6.1) are sufficient for uniqueness in the displacement problem while in the traction problem it is shown that these values may be relaxed to

$$\sigma \neq 1, \quad \mu \neq 0. \tag{6.2}$$

Turteltaub and Sternberg arrive at their conclusions by first establishing certain estimates and then relying upon Duffin's reflexion principles.

Another approach, used by Guha [1965] in dealing with the displacement problem, employs distribution theory to discuss uniqueness.

All the methods mentioned so far are applicable to isotropic media. Results valid for an anisotropic medium have recently been presented by Thompson [1969] using general theorems in the theory of partial differential equations. The displacement problem is treated in detail and it is shown what form the non-unique solutions must necessarily take.

We shall restrict our attention in this chapter to classical solutions of problems in isotropic homogeneous media, and accordingly present comparatively simple proofs of uniqueness theorems. Moreover, since we are not concerned with the structure of non-unique solutions, there is no need to employ the sophisticated arguments of Thompson [1969]. Several boundary value problems are treated, including those of mixed type. For the most part, we use Duffin's reflexion principles together with some auxiliary lemmas involving estimates, but in one of the mixed problems it is convenient to also introduce the Almansi decomposition of the biharmonic displacement. The theorems are actually established for N-dimensions; thus the half-space and the half-plane are simultaneously treated. It should be noted that Duffin formulated his reflexion principles strictly in three-dimensions. However, adaptations to two-dimensional cases are easily accomplished, and lead to analogous principles.[1] Further, results for both the half-space and the half-plane can also be deduced by applying suitable limiting procedures to the results known for the sphere and circle;[2] however, we derive such results directly in order to avoid the necessity of justifying awkward limiting operations inevitable in such procedures.

6.1 Specification of the Various Boundary Value Problems. Continuity onto the Boundary and in the Neighbourhood of Infinity

The body is assumed to occupy the region $x_1 \geqq 0$, and to have for its boundary E the hyperplane $x_1 = 0$. We let $x_1, x_2, ..., x_N$ be the cartesian coordinates. The problems considered are:

(a) the displacement components are prescribed on E, while at infinity the difference of any two displacement fields remains bounded,

(b) the traction is prescribed on E and the cartesian components of the difference between any two stress distributions is bounded at infinity,

(c) the shear stress and normal component of the displacement are prescribed on E, and the difference of any two displacement fields is bounded at infinity,

(d) the normal component of the traction and the tangential components of the displacement are prescribed on E, and the difference of any two displacement fields is bounded at infinity,

(e) the shear stress is prescribed on E, with the normal component of the displacement being given on Σ, and the normal component of the traction being specified on E/Σ. Here, Σ is a bounded subset of E, and

---

[1] See also Muskhelisvili [1953 e.g., p. 451] where similar continuation formulae are used in the solution of problems. Alternatively, the uniqueness results for the plane problems can be read off from the complete solutions presented by Muskhelisvili [1953, p. 373]. In this respect, see also Tiffen [1952].

[2] Observe that the half-space conditions are the intersection of those for the interior and exterior sphere problems. Similar remarks hold for the half-plane.

$\bar{\Sigma}$ is its closure. At infinity the difference of any two displacement fields is bounded.

Actually, the uniqueness theorems we prove permit a slightly more general behaviour at infinity than is listed here.

Let us observe once more that in isotropic elasticity the cartesian components of the displacement and stress are biharmonic provided

$$\sigma \neq 1, \quad \mu \neq 0. \tag{6.1.1}$$

Consequently, in the present chapter these conditions are always assumed to hold. This represents no loss since non-uniqueness can be easily demonstrated for values of the elasticities exceptional to (6.1.1). Now, in studying uniqueness the boundary conditions become homogeneous and therefore Duffin's reflexion principles are applicable. In problems (a)–(d) the displacement can by these principles be continued analytically across the boundary $x_1 = 0$ as a solution of the Navier equations in the whole space. After the continuations have been performed, known results for the behaviour of harmonic and biharmonic functions in the whole space can be employed in the derivation of uniqueness. A few of these results are rederived in the following three lemmas and although their proofs are already available in the literature, they are repeated here for convenience.

In problem (e) the displacement can also be continued across $x_1 = 0$ except that along the cylindrical surfaces generated by the boundary of Σ there will, in general, be singularities. For this reason we shall adopt another method of proof for (e).

Lemma 6.1.1. *Suppose the function v is biharmonic in E_N[3] and satisfies the asymptotic behaviour*

$$\lim_{x_i x_i \to \infty} v/x_i x_i = 0, \quad i = 1, 2, \ldots, N. \tag{6.1.2}$$

Then in E_N, v is a linear function of $x_1, x_2, \ldots, x_N$.

The proof begins with a mean value theorem for biharmonic functions. Let $S(r)$ denote the interior of the hypersphere whose radius is r and whose centre is at a point P in E_N. Denote the boundary of $S(r)$ by $\partial S(r)$. Then for the biharmonic function v the following mean value theorem is valid (see e.g., Bramble and Payne [1966], Duffin [1955]):

$$\Omega_N v(P) = \left(r_2^{-2} \int_{\partial S(r_2)} v \, d\Omega - r_1^{-2} \int_{\partial S(r_1)} v \, d\Omega \right) / (r_2^{-2} - r_1^{-2}), \tag{6.1.3}$$

where $d\Omega$ denotes the element of solid angle, Ω_N the surface area of the unit sphere in N-dimensions, and r_1, r_2 the arbitrary radii of concentric spheres with centre at P. Let us temporarily locate the origin at P and

---

[3] The whole N-dimensional space is denoted by E_N.

let $r_1 \to \infty$ in (6.1.3). Then, by virtue of (6.1.2) we get

$$\Omega_N v(P) = \int\limits_{\partial S(r_2)} v \, d\Omega. \tag{6.1.4}$$

But this result holds for all points P in E_N and all radii r_2, a property characterising harmonic functions. Thus, we conclude that v must be harmonic in E_N, and hence for arbitrary j and k, $\partial^2 v / \partial x_j \partial x_k$ is likewise harmonic. We now apply the harmonic mean value theorem to $\partial^2 v / \partial x_j \partial x_k$ followed by an integration by parts to obtain

$$\Omega_N \frac{\partial^2 v}{\partial x_j \partial x_k}(P) = \frac{N}{r^N} \int\limits_{S(r)} \frac{\partial^2 v}{\partial x_j \partial x_k} \, dx$$

$$= \frac{N}{r^2} \int\limits_{\partial S(r)} \frac{\partial v}{\partial x_j} x_k \, d\Omega.$$

Multiplication of the last equation by r^{N+1} and integration from $r=0$ to $r=r_1$ gives

$$\Omega_N r_1^{N+2} \frac{\partial^2 v}{\partial x_j \partial x_k}(P) = N(N+2) \int\limits_{S(r_1)} \frac{\partial v}{\partial x_j} x_k \, dx$$

$$= N(N+2) \Big\{ r_1^N \int\limits_{\partial S(r_1)} v \, n_k n_j \, d\Omega - \int\limits_{S(r_1)} v \, \delta_{jk} \, dx \Big\}. \tag{6.1.5}$$

Suppose now that ρ denotes the distance from the origin while r_1 denotes the distance from P. Then, for arbitrary ρ and r_1 (6.1.5) may be rewritten

$$\frac{\partial^2 v}{\partial x_j \partial x_k}(P) = \frac{N(N+2)}{\Omega_N} \Big\{ \int\limits_{\partial S(r_1)} \frac{v}{\rho^2} \frac{\rho^2}{r_1^2} n_k n_j \, d\Omega - \frac{1}{r_1^{N+2}} \int\limits_{S(r_1)} \frac{v}{\rho^2} \rho^2 \delta_{jk} \, dx \Big\}. \tag{6.1.6}$$

Since we are interested in large values of ρ, i.e., the limit as $\rho \to \infty$, we first choose ρ to be large but fixed and then select r_1 to be such that in $S(r)$,

$$\rho \geq r_1 \geq \rho/3.$$

In these circumstances, (6.1.6) yields

$$\left| \frac{\partial^2 v}{\partial x_j \partial x_k}(P) \right| \leq \frac{N(N+2)}{\Omega_N} \Big\{ 9\Omega_N \max_{\partial S(r_1)} \left| \frac{v}{\rho^2} \right| + \frac{9\Omega_N}{N} \max_{S(r_1)} \left| \frac{v}{\rho^2} \right| \Big\},$$

where $\max\limits_{\Sigma} |\cdot|$ indicates the maximum modulus of the value of the function over the set Σ. On letting the point P tend to infinity and on using (6.1.2) we conclude that $\partial^2 v / \partial x_j \partial x_k(P)$ vanishes at infinity. Liouville's theorem then implies that $\partial^2 v / \partial x_j \partial x_k$ vanishes identically in E_N. But j, k are arbitrary and so v must therefore be a linear function of $x_1, x_2, \ldots, x_N$ and the lemma is proved.

An easy corollary to the lemma is

Corollary 6.1.1. *If a function v is biharmonic in E_N and satisfies*

$$\lim_{x_i x_i \to \infty} v/(x_i x_i)^{\frac{1}{2}} = 0 \qquad (6.1.7)$$

then v is everywhere constant in E_N.

Lemma 6.1.2. *Suppose either the (biharmonic) cartesian components of displacement or those of stress are uniformly bounded by some constant M in $x_1 \geqq 0$. Then under these respective conditions either*

$$x_1 \left| \frac{\partial u_i}{\partial x_j}(\mathbf{x}) \right| \leqq 9M \quad \text{or} \quad x_1 \left| \frac{\partial \sigma_{ij}}{\partial x_k}(\mathbf{x}) \right| \leqq 9M. \qquad (6.1.8)$$

This lemma, due to Turteltaub and Sternberg [1967], depends upon the following version of the biharmonic mean value theorem (see, e.g., Duffin [1956]; Duffin and Noll [1958]): Let v be a function biharmonic in E_N. In the notation of Lemma 6.1.1 we then have,

$$3\Omega_N v(P) = 4 \int_{\partial S(r)} v \, d\Omega - \int_{\partial S(2r)} v \, d\Omega. \qquad (6.1.9)$$

If we now replace v by either $\partial u_i/\partial x_j$ or $\partial \sigma_{ij}/\partial x_k$, multiply by r and integrate between 0 and r, we get (after an application of the divergence theorem and the uniform bound M) the desired result (6.1.8).

Lemma 6.1.3. *Suppose that either the components of displacement or of stress possess the behaviour specified for the biharmonic function in Lemma 6.1.2. Let homogeneous conditions of type (a), (b), (c) or (d) be prescribed on $x_1 = 0$ outside some compact set Σ and, in addition, for (a) let*

$$(\lambda + 3\mu) \neq 0. \qquad (6.1.10)$$

Then under these respective conditions either

$$\rho \left| \frac{\partial u_i}{\partial x_j}(\mathbf{x}) \right| \leqq 9M \quad \text{or} \quad \rho \left| \frac{\partial \sigma_{ij}}{\partial x_k}(\mathbf{x}) \right| \leqq 9M, \quad x_1 \geqq 0$$

for $\rho = (x_i x_i)^{\frac{1}{2}}$ sufficiently large.

To prove this lemma we can make use of Duffin's reflexion principles valid if $\sigma \neq 1$, $\mu \neq 0$ in all cases except (a) where (6.1.10) must also be satisfied. Then, the displacement can be continued analytically across the boundary $x_1 = 0$ outside Σ as a solution of Navier's equation. For $|x_1|$ sufficiently large the previous lemma may be applied to obtain the desired result. For x_1 finite and $x_2^2 + x_3^2$ sufficiently large we apply the mean value theorem (6.1.9) to a sphere whose intersection with the boundary $x_1 = 0$ contains no points of Σ, the centre of the sphere lying in $x_1 \geqq 0$. The lemma is now proved as in Lemma 6.1.2.

We turn now to the consideration of uniqueness theorems.

6.2 Uniqueness of Problems (a)–(d). Corollaries for the Space E_N

We examine the question of uniqueness for problems (a)–(d), deferring until Section 6.3 the treatment of problem (e). We consider the difference v_i of two possible displacement vectors satisfying the same data and find conditions which will guarantee that v_i is either the null solution or at most a rigid body displacement. The first theorem we prove concerns the displacement boundary value problem. In various forms it has previously been established by Fredholm [1906], Knops [1965b] and Turteltaub and Sternberg [1967]. The proof we present is valid in either two or three dimensions.

Theorem 6.2.1. *There is at most one classical solution to the homogeneous isotropic elastostatic displacement boundary value problem in* $x_1 \geq 0$*, provided*

$$\sigma \neq \tfrac{3}{4}, \ 1, \quad \mu \neq 0, \tag{6.2.1}$$

and the difference v_i *of any two displacement vectors satisfies the Navier equations in* $x_1 > 0$*, together with*

$$\lim_{\mathbf{x} \to \mathbf{y}} v_i(\mathbf{x}) = 0, \quad \mathbf{y} \in E, \tag{6.2.2}$$

and

$$\lim_{\rho \to \infty} v_i(\mathbf{x})/\rho = 0, \quad x_1 \geq 0. \tag{6.2.3}$$

Conditions (6.2.1) *are also necessary. In* (6.2.3) $\rho = (x_i x_i)^{\frac{1}{2}}$.

In any actual problem condition (6.2.3) implies that at infinity any singularities in the displacement of order higher than the first must be prescribed. In order to prove the theorem we note that because (6.2.1) holds, Duffin's reflexion principle (suitably modified in two-dimensions) permits the continuation of v_i across E as a biharmonic solution of the Navier equations in E_N. Furthermore, it follows that in E_N the behaviour of v_i at infinity is governed by (6.2.3). But then, because v_i is biharmonic, Corollary 6.1.1 implies that v_i is constant everywhere in E_N. Finally, the boundary condition (6.2.2) makes this constant zero, and the sufficiency part of the theorem is proved.

To establish the necessity of (6.2.1), observe that when $\sigma = 1$, $\mu \neq 0$, a solution satisfying conditions (6.2.2) and (6.2.3) is given by $v_i = \partial \varphi / \partial x_i$, where $x_1 \geq 0$ and

$$\varphi = x_1^2 (x_i x_i - a^2)^4, \quad x_i x_i \leq a^2$$

$$= 0, \qquad\qquad x_i x_i \geq a^2.$$

When $\mu = 0$, $\sigma \neq 1$, $\tfrac{3}{4}$ any solenoidal vector vanishing on E and obeying (6.2.3) is a solution to the problem. When $\sigma = \tfrac{3}{4}$, $\mu \neq 0$ a solution in accord with the conditions of the problem is $v_i = x_1 \partial \varphi / \partial x_i$, where $\varphi(\mathbf{x})$ is any regular harmonic function in $x_1 > 0$. For instance, we could take

$$\varphi(\mathbf{x}) = [(x_1 + a)^2 + x_2^2 + \cdots x_N^2]^{-\frac{1}{2}}.$$

This completes the proof of the necessity of conditions (6.2.1).

Uniqueness in the incompressible case may be separately considered as follows. From (4.3.18) and (4.3.19) note that the displacement is still biharmonic provided $\mu \neq 0$. Thus v_i is still biharmonic and may be continued across $x_1 = 0$ as a biharmonic solution of (4.3.18) in E_N. We could now proceed as before but it is perhaps worth noting that (6.2.3) may be replaced by

$$\lim_{\rho \to \infty} v_1(\mathbf{x})/\rho^2 = 0, \qquad \lim_{\rho \to \infty} v_\alpha(\mathbf{x})/\rho = 0, \qquad \alpha = 2, \ldots, N, \qquad (6.2.4)$$

and uniqueness will continue to hold. The assertion is proved by first observing that v_α may be shown to be identically zero in $x_1 \geq 0$ exactly as indicated above. On the other hand, we conclude from Lemma 6.1.1 that v_1 is at most a linear function which by (6.2.2) must vanish on $x_1 = 0$. Also, the incompressibility condition, $\partial v_i / \partial x_i = 0$, must be satisfied. These imply that v_i must vanish identically in $x_1 \geq 0$. (The stress associated with v_i is, of course, an arbitrary hydrostatic pressure.) The entire statement of Theorem 6.2.1 is therefore proved.

The next corollary concerns an analogous result for the whole space E_N, and has been previously discussed by Muskhelisvili [1953, p. 124] and Knops [1964].

Corollary 6.2.1. *There is at most one solution to the Navier equations (2.3.2) in E_N provided the displacement uniformly approaches a prescribed value at infinity and*

$$\sigma \neq 1, \qquad \mu \neq 0. \qquad (6.2.5)$$

Duffin's reflexion principles are not required in the proof of the corollary, and hence the condition $\sigma \neq \frac{3}{4}$, essential for their validity, may be omitted. By virtue of (6.2.5), the difference v_i of any two solutions is biharmonic in E_N, while by hypothesis at infinity the behaviour of v_i is governed by

$$\lim_{\rho \to \infty} v_i = 0. \qquad (6.2.6)$$

Corollary 6.1.1 immediately restricts v_i to be constant and (6.2.6) then reduces this constant to zero, thus completing the proof of the present corollary.

Condition (6.2.5) is also necessary for uniqueness. When $\sigma = 1$, $\mu \neq 0$, we may take as a displacement satisfying the remaining hypotheses of Corollary 6.2.1, the expression

$$v_i = \partial\varphi/\partial x_i,$$

where

$$\varphi = (x_i x_i - a^2)^4, \qquad x_i x_i \leq a^2$$
$$= 0 \qquad\qquad x_i x_i \geq a^2.$$

Non-uniqueness is clearly demonstrated. When $\sigma \neq 1$, $\mu = 0$, we may as above construct a non-unique solution from any solenoidal vector vanishing uniformly at infinity.

We next consider the uniqueness of the traction boundary value problem, again in the half-space and half-plane and prove a theorem which slightly generalises a version given originally by Turteltaub and Sternberg [1967].

Theorem 6.2.2. *There is at most one stress distribution for the homogeneous isotropic elastostatic traction boundary value problem in $x_1 \geq 0$ provided the difference σ_{ij} of any two stress distributions obeys*

$$\lim_{x \to y} \sigma_{1i} = 0, \qquad y \in E, \tag{6.2.7}$$

$$\lim_{\rho \to \infty} \sigma_{1i}/\rho = 0, \qquad x_1 \geq 0, \tag{6.2.8}$$

$$\lim_{\rho \to \infty} \sigma_{\alpha\beta} = 0, \qquad x_1 \geq 0, \qquad \alpha, \beta = 2, \ldots, N \tag{6.2.9}$$

and, in addition,

$$\sigma \neq 1, \qquad \mu \neq 0. \tag{6.2.10}$$

The displacement is determined to within a rigid body motion. Conditions (6.2.10) are also necessary.

Conditions (6.2.7) imply that in any actual problem the stress field approaches continuously the surface traction which is specified everywhere on the boundary E of the region under consideration. Moreover, conditions (6.2.8) are satisfied in any actual problem if the components of stress corresponding to σ_{1i} are bounded at infinity, while (6.2.9) implies that the remaining stress components uniformly approach prescribed values at infinity.

The proof of this theorem is similar to the previous one and likewise employs Duffin's reflexion principles valid now under the hypothesis (6.2.10). The stress, σ_{ij}, can be continued across E, as a solution to the governing equations, into the whole space E_N, with condition (6.2.8) determining the asymptotic behaviour of the continued solution. Hence, we conclude that σ_{1i} and its continuation define a biharmonic function in E_N. Hence by Corollary 6.1.2 and conditions (6.2.7), (6.2.8) we may as before deduce that the stress components σ_{1i} vanish identically in $x_1 \geq 0$. The remaining stress components $\sigma_{\alpha\beta}$ therefore form an $(N-1)$-dimensional stress field in the whole sub-space E_{N-1}. They are biharmonic and so by Corollary 6.1.1 and (6.2.9) vanish identically in $x_1 \geq 0$.

The necessity of (6.2.10) can be demonstrated using the appropriate counter-examples discussed in the proof of Theorem 6.2.1, but with the displacement in each case augmented by a rigid body displacement. This establishes the present theorem in its entirety.

An immediate corollary is:[4]

Corollary 6.2.2. *For a homogeneous isotropic elastic body occupying the whole space E_N there is at most one stress distribution provided $\sigma \neq 1$, $\mu \neq 0$, and the components of stress uniformly approach prescribed values at infinity.*

The final theorem established in this section concerns the mixed boundary value problems of types (c) and (d).

Theorem 6.2.3. *Consider a homogeneous isotropic medium occupying the region $x_1 \geq 0$, and let v_i and σ_{ij} denote the difference of two displacement and stress fields, respectively. Then σ_{ij} is identically zero in $x_1 \geq 0$ and v_i is at most a rigid body translational displacement in $x_1 \geq 0$ provided*

$$\sigma \neq 1, \quad \mu \neq 0, \tag{6.2.10}$$

and the following conditions are satisfied:

$$\lim_{\rho \to \infty} v_i(\mathbf{x})/\rho = 0, \quad x_1 \geq 0 \tag{6.2.11}$$

and on the boundary either

$$\lim_{\mathbf{x} \to \mathbf{y}} v_1(\mathbf{x}) = 0, \quad \lim_{\mathbf{x} \to \mathbf{y}} \sigma_{1\alpha}(\mathbf{x}) = 0, \quad \mathbf{y} \in E, \quad \alpha = 2, \dots, N, \tag{6.2.12}$$

or

$$\lim_{\mathbf{x} \to \mathbf{y}} v_\alpha(\mathbf{x}) = 0, \quad \lim_{\mathbf{x} \to \mathbf{y}} \sigma_{11}(\mathbf{x}) = 0, \quad \mathbf{y} \in E, \quad \alpha = 2, \dots, N \tag{6.2.13}$$

is valid. Conditions (6.2.10) are also necessary.

Clearly, Theorem 6.2.3 guarantees uniqueness in problems (c) and (d). We observe again that in any actual problem condition (6.2.11) is satisfied if the displacement remains bounded at infinity, while conditions (6.2.12), (6.2.13) stipulate that the indicated components of displacement and stress must continuously approach their given surface values.

To prove Theorem 6.2.3 we again appeal to Duffin's reflexion principles, valid under (6.2.10), and then essentially repeat the proof

---

[4] See also Muskhelisvili [1953, p. 124] and Knops [1964].

of Theorem 6.2.1. In this way, we immediately deduce that v_i is constant in $x_1 \geq 0$ so that from the boundary conditions (6.2.12) we have

$$v_1(\mathbf{x})=0, \quad v_\alpha(\mathbf{x})=c_\alpha, \quad x_1 \geq 0 \qquad (6.2.14)$$

and from the boundary conditions (6.2.13) we have

$$v_1(\mathbf{x})=c_1, \quad v_\alpha(\mathbf{x})=0, \quad x_1 \geq 0, \qquad (6.2.15)$$

where c_i represents a constant.[5] The proof of the theorem is completed by employing the counter-examples of Theorem 6.2.1 to show that (6.2.10) are also necessary for uniqueness.

It is worth noting that uniqueness of the displacement in problems (c) and (d) can be insured if instead of the first of conditions (6.2.12) and (6.2.13) we require respectively

$$\lim_{\rho \to \infty} v_\alpha = 0,$$

$$\lim_{\rho \to \infty} v_1 = 0.$$

This conclusion follows at once from expressions (6.2.14), (6.2.15).

6.3. Uniqueness for the Mixed-Mixed Problem of Type (e)

For this problem we are no longer able to use reflexion principles to continue the solution across E into the whole space E_N. Instead we base the proof of uniqueness upon a complete representation formula which we shall now derive.

6.3.1 A Complete Representation of the Biharmonic Displacement in a Homogeneous Isotropic Body Occupying the Half-Space.
We concern ourselves exclusively with a homogeneous isotropic elastic body occupying the half-space $x_1 \geq 0$ and restrict the elasticities to satisfy (6.1.1) (i.e., $\sigma \neq 1$, $\mu \neq 0$). Then the displacement u_i is biharmonic and hence is completely represented by the Almansi decomposition in terms of harmonic functions $\hat{h}_i$, H_i, i.e.,

$$u_i = \hat{h}_i + x_1 H_i, \quad i = 1, 2, 3. \qquad (6.3.1)$$

We first prove that H_i is the vector gradient of a harmonic scalar. From (6.3.1) we have

$$\partial u_j/\partial x_j = \partial \hat{h}_j/\partial x_j + H_1 + x_1\, \partial H_j/\partial x_j, \qquad (6.3.2)$$

---

[5] Note that Condition (6.2.11) automatically eliminates any rigid body rotations.

and since under (6.1.1) the dilatation is harmonic, we extract from (6.3.2) the fact that

$$\partial^2 H_j/\partial x_j \, \partial x_1 = 0. \tag{6.3.3}$$

Moreover, from (6.3.1) we also obtain

$$\frac{\partial u_i}{\partial x_j} - \frac{\partial u_j}{\partial x_i} = \frac{\partial \hat{h}_i}{\partial x_j} - \frac{\partial \hat{h}_j}{\partial x_i} + \delta_{1j} H_i - \delta_{1i} H_j + x_1 \left(\frac{\partial H_i}{\partial x_j} - \frac{\partial H_j}{\partial x_i} \right); \tag{6.3.4}$$

but under (6.1.1), the expression $(\partial u_i/\partial x_j - \partial u_j/\partial x_i)$ is also harmonic, and therefore we obtain

$$\frac{\partial^2 H_i}{\partial x_j \, \partial x_1} - \frac{\partial^2 H_j}{\partial x_i \, \partial x_1} = 0. \tag{6.3.5}$$

Eq. (6.3.3) and (6.3.5) together show that $\partial H_i/\partial x_1 = \partial G/\partial x_i$ where G is harmonic in $x_1 > 0$. However, any harmonic function can be completely represented in $x_1 > 0$ as the derivative with respect to x_1 of any other harmonic function (Duffin [1955]), and hence we may write $G = \partial \hat{H}/\partial x_1$, where $\hat{H}$ is harmonic in $x_1 > 0$. Thus, we conclude that

$$H_i = \partial \hat{H}/\partial x_i + m_i(x_2, \ldots, x_N), \tag{6.3.6}$$

in which $m_i(x_2, \ldots, x_N)$ is a harmonic function of the variables $x_2, \ldots, x_N$ in $x_1 > 0$. Next, we define functions h_i by means of the relation

$$h_i = \hat{h}_i + x_1 m_i(x_2, \ldots, x_N) - \hat{H} \delta_{1i},$$

and then, clearly, h_i is harmonic in $x_1 > 0$ and expression (6.3.1) for the displacement becomes, on elimination of $\hat{h}_i$,

$$u_i = h_i + \hat{H} \delta_{1i} - x_1 \partial \hat{H}/\partial x_i. \tag{6.3.7}$$

It follows from (6.3.7) that

$$\partial u_j/\partial x_j = \partial h_j/\partial x_j \tag{6.3.8}$$

while the Navier equations (2.3.2) yield

$$-2\mu \frac{\partial^2 \hat{H}}{\partial x_i \, \partial x_1} + (\lambda + \mu) \frac{\partial^2 h_j}{\partial x_j \, \partial x_i} = 0.$$

After an integration we therefore obtain

$$(\lambda + \mu) \frac{\partial h_j}{\partial x_j} = 2\mu \frac{\partial \hat{H}}{\partial x_1} + k \tag{6.3.9}$$

for constant k. We are now in a position to obtain the required representation formula. Introduce the function H by the definition

$$H = \hat{H} + \frac{1}{2\mu} k x_1. \tag{6.3.10}$$

Clearly, H is harmonic in $x_1 > 0$ and (6.3.7) reduces to

$$u_i = h_i + H \delta_{1i} - x_1 \, \partial H / \partial x_i \qquad (6.3.11)$$

where, from (6.3.9) and (6.3.10),

$$(\lambda + \mu) \, \partial h_j / \partial x_j = 2\mu \, \partial H / \partial x_1. \qquad (6.3.12)$$

The complete representation (6.3.11), subject to (6.1.1) and (6.3.12), is the one used in the following uniqueness proof. Note that in terms of the function h_i and H the stress components are given by

$$(\lambda + \mu) \, \sigma_{ij} = 2\lambda \mu \, \frac{\partial H}{\partial x_1} \, \delta_{ij} + \mu (\lambda + \mu) \left(\frac{\partial h_i}{\partial x_j} + \frac{\partial h_j}{\partial x_i} - 2x_1 \frac{\partial^2 H}{\partial x_i \, \partial x_j} \right). \qquad (6.3.13)$$

When the material is incompressible, it follows that the displacement is still represented by (6.3.11) but that (6.3.12) is replaced by

$$6\mu \, \frac{\partial H}{\partial x_1} = \sigma_{kk}, \qquad \frac{\partial h_j}{\partial x_j} = 0, \qquad (6.3.14)$$

while the stress components are expressed by

$$\sigma_{ij} = 2\mu \, \frac{\partial H}{\partial x_1} \, \delta_{ij} + 2\mu \left(\frac{\partial h_i}{\partial x_j} + \frac{\partial h_j}{\partial x_i} - 2x_1 \frac{\partial^2 H}{\partial x_i \, \partial x_j} \right). \qquad (6.3.15)$$

6.3.2 Uniqueness in the Mixed-Mixed Problem (e).

Problem (e) corresponds to a punch problem, studied extensively in the literature (see, e.g., Green and Zerna [1968], Sneddon [1951] and Muskhelisvili [1953, p. 471]). To establish uniqueness we consider the difference of any two solutions, denoting the corresponding components of displacement and stress by v_i and σ_{ij} respectively. The homogeneous boundary conditions satisfied by these quantities are:

$$\lim_{x \to y} \sigma_{12} = 0 \qquad y \in E, \qquad (6.3.16)$$

$$\lim_{x \to y} \sigma_{13} = 0 \qquad y \in E, \qquad (6.3.17)$$

$$\lim_{x \to y} \sigma_{11} = 0 \qquad y \in E/\Sigma, \qquad (6.3.18)$$

$$\lim_{x \to y} u_1 = 0 \qquad y \in \bar{\Sigma}, \qquad (6.3.19)$$

where Σ is a bounded subset of E, and $\bar{\Sigma}$ is its closure.

We now wish to prove the following theorem:

Theorem 6.3.1. *Let v_i satisfy the Navier equations (2.3.2) in $x_1 > 0$ and the homogeneous boundary conditions (6.3.16)—(6.3.19). In addition, let*

$$\sigma \neq 1, \quad \mu \neq 0 \qquad (6.1.1)$$

and further suppose that at infinity

$$\lim_{\rho \to \infty} v_i(\mathbf{x})/\rho = 0, \quad \rho^2 = x_i x_i \qquad (6.3.20)$$

while at points of $\partial\Sigma$ the dilatation, $\partial v_j/\partial x_j$, is uniformly continuous.[6] Then v_1 and the stress components σ_{ij}, associated with v_i, vanish identically in $x_1 \geq 0$, and the remaining displacements are constant.

Theorem 6.3.1 clearly guarantees the uniqueness of the stress distribution in Problem (e).

We begin the proof of the theorem by remarking that from the traditional reflexion principles for harmonic functions, together with a lemma due to Duffin [1955, Lemma 3], it follows from (6.3.16)—(6.3.19) and (6.3.13) that

$$\partial h_1/\partial x_2 + \partial h_2/\partial x_1 = 0 \quad \text{on } E, \qquad (6.3.21)$$

$$\partial h_1/\partial x_3 + \partial h_3/\partial x_1 = 0 \quad \text{on } E, \qquad (6.3.22)$$

$$\lambda\, \partial H/\partial x_1 + (\lambda + \mu)\, \partial h_1/\partial x_1 = 0 \quad \text{on } E/\Sigma, \qquad (6.3.23)$$

$$h_1 + H = 0 \quad \text{on } \Sigma. \qquad (6.3.24)$$

Further, assuming (6.3.20) to be satisfied we may use the mean value theorem (compare Lemma 6.1.3) to deduce that the dilatation and stress vanish as $\rho \to \infty$. By (6.3.8) and (6.3.12) this implies that $\partial H/\partial x_1$ also vanishes as $\rho \to \infty$ and that

$$\lim_{\rho \to \infty} \rho\, \frac{\partial^2 H}{\partial x_i\, \partial x_1} = 0. \qquad (6.3.25)$$

Expression (6.3.13) then enables us to conclude that $\partial h_1/\partial x_\alpha + \partial h_\alpha/\partial x_1$ ($\alpha = 2 \ldots N$) also vanishes as $\rho \to \infty$. But $\partial h_1/\partial x_\alpha + \partial h_\alpha/\partial x_1$ is harmonic in $x_1 > 0$ and from (6.3.21) and (6.3.22) vanishes on $x_1 = 0$. Thus, by a well known theorem in potential theory, we have

$$\frac{\partial h_1}{\partial x_\alpha} + \frac{\partial h_\alpha}{\partial x_1} \equiv 0 \quad \text{in } x_1 \geq 0. \qquad (6.3.26)$$

A differentiation with respect to x_α, followed by a sum from $\alpha = 2$ to $\alpha = N$ and use of the harmonicity of h_1 then produces

$$2\, \frac{\partial^2 h_1}{\partial x_1\, \partial x_1} - \frac{\partial^2 h_j}{\partial x_j\, \partial x_1} \equiv 0 \quad \text{in } x_1 > 0. \qquad (6.3.27)$$

---

[6] In the incompressible case, the last condition is replaced by the uniform continuity at points of $\partial\Sigma$ of the derivative with respect to x_1 of the hydrostatic pressure.

An elimination of $\partial h_j/\partial x_j$ between (6.3.27) and (6.3.12) and a subsequent integration results in

$$(\lambda+\mu)\frac{\partial h_1}{\partial x_1}-\mu\frac{\partial H}{\partial x_1}=g(x_2,\ldots,x_N)\quad\text{in }x_1>0, \qquad (6.3.28)$$

for a harmonic function g. However, by (6.3.13) and (6.3.20) we have

$$(\lambda+\mu)\sigma_{11}=2\mu\left[(\lambda+\mu)\frac{\partial H}{\partial x_1}+g-(\lambda+\mu)x_1\frac{\partial^2 H}{\partial x_1\,\partial x_1}\right],$$

and so from the vanishing at infinity of σ_{11} and (6.3.25) we conclude that g must be identically zero; thus

$$(\lambda+\mu)\frac{\partial h_1}{\partial x_1}-\mu\frac{\partial H}{\partial x_1}=0\quad\text{in }x_1>0. \qquad (6.3.29)$$

Combining the last result with (6.3.23) next shows that provided $\lambda+\mu\neq0$ (i.e., the material is compressible),

$$\lim_{x\to y}\frac{\partial h_1}{\partial x_1}=\lim_{x\to y}\frac{\partial H}{\partial x_1}=0,\quad y\in E/\Sigma. \qquad (6.3.30)$$

However, from the reflexion principles for harmonic functions we know that the limiting values are actually attained on E/Σ.

Again, reflexion principles show that h_1+H is harmonic on Σ so that by differentiation we may deduce

$$\frac{\partial^2 h_1}{\partial x_1\,\partial x_1}+\frac{\partial^2 H}{\partial x_1\,\partial x_1}=0\quad\text{on }\Sigma \qquad (6.3.31)$$

and then a similar argument to that just used may be combined with (6.3.29) (or rather the primitive form before integration with respect to x_1) to show that provided $\lambda+2\mu\neq0$, $\mu\neq0$,

$$\frac{\partial^2 h_1}{\partial x_1\,\partial x_1}=\frac{\partial^2 H}{\partial x_1\,\partial x_1}=0\quad\text{on }\Sigma. \qquad (6.3.32)$$

Since $\partial v_j/\partial x_j$ has been supposed uniformly continuous at points on $\partial\Sigma$, we gather from (6.3.8) and (6.3.12) that provided $\lambda+\mu\neq0$ this property extends also to $\partial H/\partial x_1$. Moreover, $\partial H/\partial x_1$ vanishes at infinity. Consequently, from (6.3.30), (6.3.32) and the maximum principle of Hopf we conclude that $\partial H/\partial x_1$ vanishes identically in $x_1\geq0$. We immediately deduce from (6.3.29) that $\partial h_1/\partial x_1$ vanishes identically in $x_1>0$, and on recalling the conditions on $\partial h_1/\partial x_1$ along E we conclude that this function also vanishes identically in $x_1\geq0$. From (6.3.13) it then follows that

$$\sigma_{1j}\equiv0\quad\text{in }x_1\geq0. \qquad (6.3.33)$$

Moreover, (6.3.8), (6.3.12) and the vanishing of $\partial H/\partial x_1$ next show that the dilatation also vanishes in $x_1 \geqq 0$. Thus, the remaining stress components $\sigma_{\alpha\beta}$ satisfy

$$\left.\begin{array}{c} \dfrac{\partial \sigma_{\alpha\beta}}{\partial x_\beta} = 0, \\[3mm] \sigma_{\alpha\beta} = \mu\left(\dfrac{\partial v_\alpha}{\partial x_\beta} + \dfrac{\partial v_\beta}{\partial x_\alpha}\right) \end{array}\right\} \quad \text{in } x_1 \geqq 0, \qquad (6.3.34)$$

showing that v_α and $\sigma_{\alpha\beta}$ are biharmonic functions in E_{N-1}. Corollary (6.1.1) then guarantees that v_α is constant in $x_1 \geqq 0$.

Finally, observe that in $x_1 \geqq 0$, v_1 reduces to a harmonic function of the variables $x_2, x_3 \ldots x_N$, and hence must be constant in $x_1 \geqq 0$ by virtue of known results in potential theory and the prescribed asymptotic behaviour (6.3.20). The boundary condition (6.3.19) then shows this constant is zero in $x_1 \geqq 0$, and so the theorem is proved.

It is worth remarking that condition (6.3.20) may be replaced by

$$\lim_{\rho \to \infty} \sigma_{ij}(\mathbf{x})/\rho = 0, \qquad (6.3.35)$$

in which case the above proof indicates that the stress distribution σ_{ij} vanishes identically in $x_1 \geqq 0$, the displacement therefore being a rigid body one.

For the incompressible case, (i.e., $\lambda + \mu = 0, \mu \neq 0$) we must use formulae (6.3.14) and (6.3.15) in the representation of v_i. The boundary conditions (6.3.21), (6.3.22), (6.3.24) remain unaltered while (6.3.23) becomes

$$\frac{\partial H}{\partial x_1} + \frac{\partial h_1}{\partial x_1} = 0 \quad \text{on } E/\Sigma. \qquad (6.3.36)$$

Eq. (6.3.25) and (6.3.26) follow as before, but because of condition (6.3.14)$_2$, Eq. (6.3.27) is now replaced by

$$\frac{\partial^2 h_1}{\partial x_1 \, \partial x_1} \equiv 0 \quad \text{in } x_1 > 0. \qquad (6.3.37)$$

Harmonic continuation formulae show in fact that (6.3.37) holds on $x_1 = 0$. Further, from (6.3.11) we get

$$\frac{\partial v_1}{\partial x_1} = \frac{\partial h_1}{\partial x_1} - x_1 \frac{\partial^2 H}{\partial x_1 \, \partial x_1}$$

and hence, because of (6.3.20) and (6.3.25), we see that $\partial h_1/\partial x_1$ vanishes uniformly as $\rho \to \infty$. From the harmonic maximum principle, together with (6.3.37), we conclude that $\partial h_1/\partial x_1$ vanishes identically in $x_1 \geqq 0$.

Thus, (6.3.36) may be written

$$\partial H/\partial x_1 = 0 \qquad \text{on } E/\Sigma, \tag{6.3.38}$$

and so by applying reflexion principles and differentiating we get

$$\partial^3 H/\partial x_1^3 = 0 \qquad \text{on } E/\Sigma. \tag{6.3.39}$$

Eq. (6.3.31) is still valid, showing with (6.3.37) that

$$\frac{\partial^2 H}{\partial x_1^2} = 0 \qquad \text{on } \Sigma. \tag{6.3.40}$$

On supposing that $\partial^2 H/\partial x_1^2$ (i.e., the derivative with respect to x_1 of the hydrostatic pressure) is uniformly continuous at points of $\partial\Sigma$, we may conclude as before that $\partial^2 H/\partial x_1^2$ vanishes identically in $x_1 \geq 0$ and therefore that

$$H = x_1 g(x_2, \ldots, x_N) + f(x_2, \ldots, x_N), \tag{6.3.41}$$

where g and f are arbitrary harmonic functions of the indicated variables. It follows from (6.3.11) that we may now write

$$v_1 = h_1 + f. \tag{6.3.42}$$

The last equation shows that v_1 is a harmonic function of $x_2, \ldots, x_N$ in $x_1 \geq 0$, and hence by (6.3.20) is constant in $x_1 \geq 0$. However, in agreement with (6.3.19), this constant vanishes. A differentiation of (6.3.41) together with (6.3.38) imply that g vanishes identically and hence that $\partial H/\partial x_1 \equiv 0$ in $x_1 \geq 0$. The Navier equations then imply that $v_2, \ldots, v_N$ are harmonic in $x_1 > 0$. But v_1 vanishes in $x_1 \geq 0$ and so (6.3.10), (6.3.17) yield

$$\frac{\partial v_\alpha}{\partial x_1} = 0 \qquad \text{on } \Sigma.$$

The harmonic maximum principle then confirms that v_α are constant in $x_1 \geq 0$ and the proof is complete.

When the field behaviour at infinity is specified by (6.3.35) a similar proof shows uniqueness continues to hold. However, it suffices now to require the uniform continuity onto $\partial\Sigma$ of the hydrostatic pressure, and not its derivative, since then (6.3.38) and (6.3.40) are adequate to prove that it vanishes identically in $x_1 \geq 0$. Together with the vanishing of $\partial h_1/\partial x_1$ in $x_1 \geq 0$ this implies

$$\sigma_{1j} = 0$$

$$\sigma_{\alpha\alpha} = 0 \qquad \alpha = 2, \ldots, N$$

and from the stress equilibrium equations we immediately obtain the harmonicity of $\sigma_{\alpha\beta}$ in E_{N-1}. Uniqueness now follows from Corollary 6.2.2.

We note finally that conditions (6.1.1) may be proved necessary by means of methods used previously.

Similar techniques could be applied to establish uniqueness for the problem in which the tangential displacements are prescribed on $x_1 = 0$, the displacement u_1 is prescribed over $\bar{\Sigma}$ and the normal component of traction on E/Σ. Uniqueness follows in a reasonably straightforward manner if $\sigma \neq \frac{3}{4}, 1, \mu \neq 0$. A finer analysis would show that $\sigma = \frac{3}{4}$ is not an exceptional value. Other boundary value problems of mixed-mixed type could also be treated using the decomposition (6.3.11). However, in the case in which the displacement components are prescribed on $\bar{\Sigma}$ and the tractions on E/Σ, one encounters difficulties in applying (6.3.11) and it is not known for what extended range of values of Poisson's ratio uniqueness holds.

Chapter 7

Miscellaneous Boundary Value Problems

In the previous sections the uniqueness question for the most important standard problems of linear anisotropic elastostatics has been studied. There remain a number of problems of non-standard type which have attracted considerable interest in the recent literature. We devote this chapter to a discussion of the uniqueness question for some of these non-standard problems. While many such problems may be physically interesting, often they are devoid of such interest, and serve at best an auxiliary role; for example, the Cauchy problem played an important part in the proof of Theorem 4.1.3, concerned with the uniqueness of the displacement boundary value problem for non-homogeneous anisotropic elasticity. In this chapter, we do not always present proofs of theorems with apparently little or no physical significance, but instead refer the reader to the paper in which the proof may be found.

We begin with problems associated with a sphere and continue by proving uniqueness in the Cauchy problem for a homogeneous isotropic elastic body in equilibrium, a result originally derived by Almansi [1907]. The second part of the chapter is an account of uniqueness in several problems characterised either by "ambiguous" conditions i.e., the Signorini problem,[1] or by related variational inequalities.[2]

In this chapter, unless otherwise indicated, we deal only with classical solutions, even though generalizations to weak solutions can frequently easily be made.

7.1 Problems for a Sphere

Diaz and Payne [1958] considered a homogeneous isotropic elastic body (interior of a sphere) and proved that the prescription of normal components of the displacement and traction over the entire surface is sufficient for the unique determination of all stress components at the centre of the sphere, provided $\mu \neq 0$ and $\sigma \neq -\frac{7}{5}, \frac{3}{5}, \frac{2}{3}, 1$. Subsequently, Bramble and Payne [1961b] examined the same spherical solid but

---

[1] See Signorini [1959a, b], Fichera [1963a, b].
[2] See Lions and Stampacchia [1967] and Brezis and Stampacchia [1968].

with the normal components of the displacement, traction and rotation prescribed on the surface. By means of an appropriate decomposition formula uniqueness was established provided $\mu \neq 0$ and

$$\sigma \neq (N-2)(2N-1)^{-1}, \tag{7.1.1}$$

where N is any positive integer. When equality holds in (7.1.1) for some particular N, it was further shown that the displacement is uniquely determined to within a function of the form

$$\frac{x_i \theta}{3} - \frac{r^3}{6} \frac{\partial \theta}{\partial x_i} + a^2 \frac{N-2}{6N} \frac{\partial \theta}{\partial x_i}, \quad r^2 = x_i x_i, \tag{7.1.2}$$

in which θ is an arbitrary harmonic polynomial of degree N and a is the radius of the sphere. Observe that (7.1.1) yields an infinite set of eigenvalues for σ, all of which lie in the physically interesting range $-1 < \sigma \leq \frac{1}{2}$.

Bramble and Payne [1961b] also proved that in the corresponding exterior problem, under suitable asymptotic behaviour at infinity, e.g., $u_i = O(r^{-1})$, $r \to \infty$, there is uniqueness provided $\mu \neq 0$ and

$$\sigma \neq \frac{1}{2} + \frac{3}{2}(2M+1)^{-1} \tag{7.1.3}$$

for any positive integer M. The values of σ for which equality holds in (7.1.3) lie in the range $\frac{1}{2} \leq \sigma \leq 1$, demonstrating possible failure of uniqueness at values of σ outside the range of physical interest.

Diaz and Payne [1958, 1963] considered a second problem also associated with the homogeneous isotropic elastic sphere. Here it was shown that the knowledge of the surface values of the tangential components of the displacement and of the traction is not sufficient to determine all the stress components at the centre of the sphere. As they mention, an arbitrary hydrostatic pressure can obviously be added without disturbing the boundary data.

This conclusion led Bramble and Payne [1961b] to consider the uniqueness aspects of the problem. They found that specifying the surface values of the tangential components of displacement and traction guaranteed, for the interior problem, a unique solution to within an arbitrary hydrostatic pressure provided condition (7.1.1) is met. For the exterior problem (e.g. with imposition of the asymptotic behaviour, $u_i = O(r^{-1}), r \to \infty$), the solution is unique to within a term of the form $C \operatorname{grad}(r^{-1})$, for an arbitrary constant C, provided (7.1.3) holds.

Note that in general a solution to the above problem fails to exist unless a compatibility relationship is prescribed for the data. See Bramble and Payne [1961b].

7.2 The Cauchy Problem for Isotropic Elastostatics

We have already encountered (see Sections 4.1.1, 4.4.2) the Cauchy problem in its strict sense, i.e., with the displacement and its normal derivative given over a portion of the bounding surface. In this section we intend dealing with the more general problem in which the specified Cauchy data are the values of the displacement and traction over an arbitrary portion of the surface. In fact, we prove the following theorem,[3] which is an extension of the original result due to Almansi [1907] for a homogeneous isotropic elastic body.

Theorem 7.2.1. *For a nonhomogeneous anisotropic elastic body occupying a region B with the displacement and traction given on the same portion ∂B* of the surface ∂B, there is at most one solution twice continuously differentiable in B ∪ ∂B*, provided the elasticities are analytic in B and satisfy at each point of B ∪ ∂B* the ellipticity condition*

$$\det |c_{ijkl}\, \xi_j \xi_l| \neq 0 \quad \text{for all} \quad \xi_i \neq 0. \tag{7.2.1}$$

The two-dimensional version of this theorem for homogeneous isotropic elasticity follows easily from properties of analytic functions and was proved, for instance, by Muskhelisvili [1953, p. 133].[4]

To prove Theorem 7.2.1, let v_i denote the difference of two possible solutions to the problem and then observe that the vanishing of v_i on ∂B* implies that

$$\frac{\partial v_i}{\partial x_j} = \lambda_i n_j \quad \text{on} \quad \partial B^*, \tag{7.2.2}$$

where n_j are the components of the unit outward normal on ∂B* and the λ_i are functions of position. Because the traction also vanishes on ∂B*, we then have

$$n_j c_{ijkl} \lambda_k n_l = 0 \quad \text{on} \quad \partial B^*, \tag{7.2.3}$$

so that from (7.2.1) we may conclude that $\lambda_i = 0$ on ∂B* and hence that $\partial v_i/\partial n = 0$ on ∂B*. Condition (7.2.1) states, in addition, that all characteristic surfaces are imaginary, so that, in particular, ∂B* is non-characteristic. Uniqueness then follows from Holmgren's theorem.

Various counter-examples may be constructed indicating that violation of condition (7.2.1) leads to non-uniqueness. For instance, in the homogeneous isotropic case, when the elasticities satisfy $\mu \neq 0$,

---

[3] The proof of Theorem 7.2.1 together with the basic idea for the general counter-example following the theorem is due to Dr. M. Hayes (private communication).

[4] Muskhelisvili [1953 p. 132] also proved a similar result, again based upon properties of analytic functions, that if an arbitrary small portion of a plane homogeneous isotropic elastic body is stress free, then the entire body is likewise stress free.

$\sigma=1$, we may merely repeat a previous construction (see e.g., p. 28) to obtain the required counter-example. For a more general case, let ∂B^* be a portion of a spherical surface of radius a, whose centre, located at the origin of coordinates, does not lie in B. We set

$$c_{ijkl}=(B_{ij}x_lA_k)/(x_px_pA_qA_q), \tag{7.2.4}$$

where A_m is constant and B_{ij} is given by

$$B_{ij}=e_{jpq}\frac{\partial v_{ip}}{\partial x_q} \tag{7.2.5}$$

with $v_{ij}(=v_{ij}(\mathbf{x}))$ analytic and chosen to make $B_{ij}x_j=0$ on ∂B^*. Then,

$$\det|c_{ijkl}x_jx_l|=\det|B_{ij}x_jA_k|=0 \quad \text{on} \quad \partial B^*. \tag{7.2.6}$$

Therefore, ∂B^* is a characteristic surface and (7.2.1) is violated. We now consider

$$u_i=A_i(x_px_p-a^2). \tag{7.2.7}$$

The displacement (7.2.7) clearly vanishes on ∂B^*, while for the associated traction we have

$$n_jc_{ijkl}\frac{\partial u_k}{\partial x_l}=\frac{2}{a}c_{ijkl}x_jA_kx_l$$

$$=\frac{2}{a}B_{ij}x_j,$$

which also vanishes on ∂B^*. Moreover, since

$$\frac{\partial}{\partial x_j}\left(c_{ijkl}\frac{\partial u_k}{\partial x_l}\right)=\frac{\partial B_{ij}}{\partial x_j},$$

we gather from (7.2.5) that the equilibrium equations are satisfied. Thus, the displacement (7.2.7) represents a non-unique solution to the Cauchy problem in which the elasticities are given by (7.2.4). In this sense, therefore, the conditions of Theorem 7.2.1 are also necessary for uniqueness.

We remark that the Cauchy problem defined in Theorem 7.2.1 finds applications in geophysics and geology.

7.3 The Signorini Problem.
Other Problems with Ambiguous Conditions

We wish now to study a problem first proposed by Signorini [1959a, b]. The elastic anisotropic body occupies a region B in the half-plane $x_1>0$ and initially a portion Σ of the surface ∂B lies on the rigid perfectly lubricated surface $x_1=0$. The elastic body, assumed to be under the influence of gravity, is now subjected to body or surface forces. In the resulting deformation points which were originally on Σ may not remain in contact with the rigid plane. Let P be a point which remains in contact with surface $x_1=0$. Then at P the following conditions hold:

$$\sigma_{12}=\sigma_{13}=0,$$

$$u_1=0, \quad \sigma_{11}\leqq0,$$
(7.3.1a)

where the displacement u_i is related to the stress components σ_{ij} through (2.1.10), i.e.,

$$\sigma_{ij}=c_{ijkl}\frac{\partial u_k}{\partial x_l}, \quad c_{ijkl}=c_{klij}.$$

On the other hand, under the assumptions of the linear theory, for a point Q, originally on Σ, which has pulled away from the plane $x_1=0$, boundary conditions of the following type are valid:

$$\sigma_{12}=\sigma_{13}=0,$$

$$\sigma_{11}=0, \quad u_1\leqq0.$$
(7.3.1b)

The corresponding boundary value problem is therefore ambiguous—at points of Σ either (7.3.1a) or (7.3.1b) holds, but at a given point it is not known a priori which one applies. Along $\partial B/\Sigma$ boundary conditions of standard type are assumed.

A similar problem arises when a bounded elastic body is acted on by a number of rigid dies.

The preceding remarks lead to the following generalized formulation of the *Signorini problem* (S): Consider a bounded anisotropic elastic body, part of whose surface ∂B is the subset Σ. The *problem S* then consists in *the determination of the stress field* σ_{ij}, *satisfying the equation*

$$\frac{\partial}{\partial x_j}\left(c_{ijkl}(\mathbf{x})\frac{\partial u_k}{\partial x_l}\right)+\rho F_i=0 \quad \text{in } B$$
(7.3.2)

and the data

$$\sigma_{ij}n_jn_k-\sigma_{kj}n_jn_i=f_{ik} \quad \text{on } \partial B$$

$$\sigma_{ij}n_in_j \qquad\qquad =g \quad \text{on } \partial B/\Sigma,$$
(7.3.3)

together with the additional hypotheses that at every point of $\bar{\Sigma}$ *either*

$$u_in_i=f, \quad \sigma_{ij}n_in_j\geqq0,$$
(7.3.4a)

or

$$u_in_i>f, \quad \sigma_{ij}n_in_j=0.$$
(7.3.4b)

Here, as usual, n_i denotes the component of the unit outward normal on ∂B, and the elasticities are assumed to be uniformly positive definite. It is not known beforehand which of the inequalities (7.3.4a), (7.3.4b) prevails at a given point of Σ. Note that the signs of the inequalities in (7.3.4a) and (7.3.4b) are the reverse of those in (7.3.1a) and (7.3.1b) in conformity with the notation introduced by Fichera [1963a].

Signorini [1959a, b] showed that, provided a solution of class $C^1(\bar{B})$ exists, it minimizes the potential energy in a non-linear class of admissible displacements. Fichera [1963b][5] established that, in fact, solutions of class $C^1(\bar{B})$ cannot in general be expected to exist. Nevertheless, by enlarging the class of admissible displacements and properly restating the problem S, he was able, by means of an extension of the techniques of the calculus of variations, to derive necessary and sufficient conditions on the data (loads) for both existence and uniqueness. With his assumptions on the elasticities, Fichera demonstrated that the solution is in the class $C^2(B)$ and in the class $C^1(B \cup A)$, where A is any subset of $\partial B/\Sigma$ or of the part Σ on which $u_i n_i = f$.

Since the advent of Fichera's work, considerable attention has been devoted in the literature to ambiguous problems in various contexts. Several of these results were presented in the articles by Hill [1967], Lewy [1967], Lions and Stampacchia [1967], Brezis and Stampacchia [1968], Nitsche [1969], Lewy and Stampacchia [1969] and Sewell [1969]. However, we confine our attention here solely to questions of uniqueness. We begin our treatment by first establishing the following theorem.

Theorem 7.3.1. *For a bounded region B, the Signorini problem S has at most one classical solution (i.e., $u_i \in C^2(B) \cup C^0(\bar{B}) \cup H^1(B)$),[6] provided the elasticities satisfy the positive-definiteness conditions* (2.4.1).

As before we assume the existence of two solutions u_i^1 and u_i^2 and set $v_i = u_i^1 - u_i^2$. Thus v_i satisfies (7.3.2) and (7.3.3) with F_i and g both identically zero. From the divergence theorem it follows that

$$0 = \int_B v_i \frac{\partial}{\partial x_j}(c_{ijkl}) \frac{\partial v_k}{\partial x_l} dx = \int_{\partial B} \sigma_{ij} n_j v_i \, dS - \int_B c_{ijkl} \frac{\partial v_k}{\partial x_l} \frac{\partial v_i}{\partial x_j} dx$$

$$= \int_\Sigma \sigma_{ij} n_i n_j v_k n_k \, dS - \int_B c_{ijkl} \frac{\partial v_k}{\partial x_l} \frac{\partial v_i}{\partial x_j} dx. \tag{7.3.5}$$

---

[5] See also the survey paper by Fichera [1964].

[6] The (Sobolev) space $H^n(B)$ is the functional completion of $C^\infty(B)$ in the Hilbert norm

$$\sum_{i=0}^n \int_B D^i u \, D^i u \, dx,$$

where

$$D^i = \frac{\partial^i}{\partial x_1^{\alpha_1} \partial x^{\alpha_2} \dots \partial x^{\alpha_n}}, \qquad \alpha_1 + \alpha_2 + \dots + \alpha_n = i.$$

Here we understand that σ_{ij} is defined in terms of v_i according to (2.1.10). We look now at the integrand of the first integral on the right. At a point P_1 on Σ we assume without loss that $u_i^1 n_i \leq u_i^2 n_i$; then one of three possibilities arises; i.e., either

$$\text{i)} \quad u_i^1 n_i > f, \quad u_i^2 n_i > f,$$

$$\text{ii)} \quad u_i^1 n_i = f, \quad u_i^2 n_i > f, \tag{7.3.6}$$

$$\text{iii)} \quad u_i^1 n_i = f, \quad u_i^2 n_i = f.$$

It follows from (7.3.4a) and (7.3.4b) that in case i) $\sigma_{ij} n_i n_j = 0$, in case ii) $\sigma_{ij} n_i n_j \geq 0$ and in case iii) $v_i n_i = 0$. Thus in all three cases it follows that $\sigma_{ij} n_i n_j n_k v_k \leq 0$ at the point P_1. Obviously the roles of u_i^1 and u_i^2 could be interchanged in the assumed inequality at any point of Σ and the same result would follow. Hence the integrand of the integral over Σ is non-positive, and (7.3.5), therefore, yields

$$\int_B c_{ijkl} \frac{\partial v_k}{\partial x_l} \frac{\partial v_i}{\partial x_j} dx \leq 0, \tag{7.3.7}$$

from which uniqueness up to a possible rigid body motion may be concluded.

We should perhaps remark that a necessary condition for the existence of a solution to (7.3.2) and (7.3.3) is the compatibility condition

$$\int_B \rho F_i dx + \int_{\partial B} f_{ik} n_k dS + \int_{\partial B/\Sigma} g n_i dS + \int_{\Sigma} \sigma_{kj} n_j n_k n_i dS = 0. \tag{7.3.8}$$

Clearly the same method establishes uniqueness when standard boundary conditions are prescribed on $\partial B/\Sigma$ and when either tangential components of the traction or tangential components of the displacement are prescribed on Σ together with the ambiguous conditions (7.3.4a) or (7.3.4b). Moreover, the same approach successfully proves uniqueness when the ambiguous conditions (7.3.4a) and (7.3.4b) are replaced by the requirement that at points of Σ either

$$\sigma_{ij} n_i n_j + \alpha(u_i n_i - h) = f, \quad u_i n_i \geq h, \tag{7.3.4a'}$$

or

$$\sigma_{ij} n_i n_j = f, \quad u_i n_i < h \tag{7.3.4b'}$$

is valid, where $\alpha \geq 0$, and h and f are assigned data. Indeed, if we denote by S' the problem of finding the solution of (7.3.2) and (7.3.3) subject to the alternative conditions (7.3.4a') or (7.3.4b'), we have

Theorem 7.3.2. *For a bounded region B, the problem S' has at most one classical solution (i.e., $u_i \in C^2(B) \cup C^0(\bar{B}) \cup H^1(B)$), provided the elasticities satisfy the positive-definiteness condition (2.4.1).*

Conditions of the type (7.3.4a') and (7.3.4b') are possibly appropriate to an elastic body (B) supported along Σ on an elastic foundation.

Another type of ambiguous problem is provided by the following situation: let us suppose that the equations of elasticity are satisfied at those points in a region B where the strain energy density is less than some constant N, while at no point in B can the strain energy density assume a value higher than N. This problem is obviously reminiscent of that occurring in the theory of elastic-plastic torsion, one for which the questions of existence and uniqueness have been studied for certain geometries by Annin [1965].[7] A typical formulation of our problem is: *for a bounded region B determine a function* $u_i \in C^1(B) \cup H_0^2(B)$ [8] *such that at any point of B either*

$$\frac{\partial}{\partial x_j}\left[c_{ijkl}\frac{\partial u_k}{\partial x_l}\right] + \rho F_i = 0 \quad \text{and} \quad c_{ijkl}\frac{\partial u_i}{\partial x_j}\frac{\partial u_k}{\partial x_l} < N \qquad (7.3.9\,\text{a})$$

or

$$c_{ijkl}\frac{\partial u_i}{\partial x_j}\frac{\partial u_k}{\partial x_l} \equiv N \qquad (7.3.9\,\text{b})$$

are to hold, while

$$u_i = 0 \quad \text{on} \quad \partial B. \qquad (7.3.10)$$

It may be proved that the function u_i, if it exists, satisfies a variational inequality.[9] Explicitly, let $\mathcal{N}$ denote the space of functions φ_i given by

$$\mathcal{N} = \left\{\varphi_i \in H_0^1(B); \ c_{ijkl}\frac{\partial \varphi_i}{\partial x_j}\frac{\partial \varphi_k}{\partial x_l} \leq N\right\}. \qquad (7.3.11)$$

Then the vector function $u_i \in \mathcal{N}$ satisfying

$$-\int_B \rho F_i(\varphi_i - u_i)\,dx + \int_B c_{ijkl}\frac{\partial u_k}{\partial x_l}\left(\frac{\partial \varphi_i}{\partial x_j} - \frac{\partial u_i}{\partial x_j}\right)dx \leq 0, \qquad (7.3.12)$$

for all $\varphi_i \in \mathcal{N}$ is the one required.

The existence and uniqueness of the function u_i in problems of this type have been examined by Lions and Stampacchia [1967], and the regularity of u_i was subsequently treated by Brezis and Stampacchia [1968]. Unfortunately, the maximum principle is not satisfied by the equations of elasticity and so the techniques of the latter paper do not directly apply to our problem. However, in the event that every solution of (7.3.12) belongs to the class $C^1(B) \cup H_0^2(B)$, then (7.3.12) may be

---

[7] See also Lanchon and Duvaut [1967] and Ting [1967].
[8] Functions belonging to $H_0^m(B)$ are those in $H^m(B)$ having compact support in B.
[9] See, e.g., Brezis and Stampacchia [1968].

rewritten

$$- \int_B \left[\rho F_i + \frac{\partial}{\partial x_j} \left(c_{ijkl} \frac{\partial u_k}{\partial x_l} \right) \right] [\varphi_i - u_i] \, dx \leq 0, \qquad \forall \varphi_i \in \mathcal{N}. \quad (7.3.13)$$

Standard variational arguments then show that the solution of (7.3.13) is precisely the desired solution to (7.3.9) and (7.3.10). What is more, solutions of (7.3.13) belonging to $C^1(B) \cup H_0^2(B)$ are unique, as the next theorem proves.

Theorem 7.3.3. *For a bounded region there is at most one solution $u_i \in C^1(B) \cup H_0^2(B)$ of the variational inequality (7.3.13) provided the elasticities satisfy the positive-definiteness condition (2.4.1).*

The proof of this theorem is straightforward. As usual, we assume the existence of two solutions u_i^1 and u_i^2. Then (7.3.12) holds for $u_i = u_i^1$ and any $\varphi_i \in \mathcal{N}$. We choose $\varphi_i = u_i^2$. Similarly, for $u_i = u_i^2$ we choose $\varphi_i = u_i^1$. This leads to

$$- \int_B \left[\rho F_i + \frac{\partial}{\partial x_j} \left\{ c_{ijkl} \frac{\partial u_k^1}{\partial x_l} \right\} \right] (u_i^2 - u_i^1) \, dx \leq 0 \qquad (7.3.14)$$

and

$$- \int_B \left[\rho F_i + \frac{\partial}{\partial x_j} \left\{ c_{ijkl} \frac{\partial u_k^2}{\partial x_l} \right\} \right] (u_i^1 - u_i^2) \, dx \leq 0. \qquad (7.3.15)$$

Adding (7.3.14) and (7.3.15) we obtain at once

$$\int_B \frac{\partial}{\partial x_j} \left[c_{ijkl} \frac{\partial}{\partial x_l} (u_k^1 - u_k^2) \right] (u_i^1 - u_i^2) \, dx \geq 0. \qquad (7.3.16)$$

An integration by parts now yields

$$\int_B c_{ijkl} \frac{\partial}{\partial x_l} (u_k^1 - u_k^2) \frac{\partial}{\partial x_j} (u_i^1 - u_i^2) \, dx \leq 0. \qquad (7.3.17)$$

But the c_{ijkl} are assumed to satisfy (2.4.1) which implies that $(u_i^1 - u_i^2) \equiv 0$, and so the theorem is proved[10].

Exactly the same arguments as those just used establish uniqueness under several other conditions. For instance, the previous proof may be repeated with the positive-definiteness of the elasticities replaced by any condition implying uniqueness in the standard displacement boundary value problem. Again, instead of boundary conditions (7.3.10), any standard homogeneous or non-homogeneous boundary condition may be used and uniqueness is obtained, provided only that now the variational

---

[10] Note that uniqueness of solutions of (7.3.12) in $H_0^1(B)$ could be proved in a similar way.

form (7.3.13) is assumed with the functions φ_i and u_i in appropriate spaces. Nor are the precise ambiguous conditions (7.3.9) essential in the proof of the theorem. They are replaceable, for example, by: *at every point of B either*

$$\frac{\partial}{\partial x_j}\left(c_{ijkl}\frac{\partial u_k}{\partial x_l}\right)+\rho F_i=0, \quad \left(\frac{\partial u_i}{\partial x_j}+\frac{\partial u_j}{\partial x_i}\right)\left(\frac{\partial u_i}{\partial x_j}+\frac{\partial u_j}{\partial x_i}\right)<N, \quad (7.3.9\,\text{a}')$$

or

$$\left(\frac{\partial u_i}{\partial x_j}+\frac{\partial u_j}{\partial x_i}\right)\left(\frac{\partial u_i}{\partial x_j}+\frac{\partial u_j}{\partial x_i}\right)\equiv N. \qquad (7.3.9\,\text{b}')$$

Another possible set of ambiguous conditions is: *at every point of B either*

$$\frac{\partial}{\partial x_j}\left(c_{ijkl}\frac{\partial u_k}{\partial x_l}\right)+\rho F_i=0, \quad u_i u_i<N, \qquad (7.3.9\,\text{a}'')$$

or

$$u_i u_i\equiv N. \qquad (7.3.9\,\text{b}'')$$

Characterization of solutions of such ambiguous problems as solutions of variational inequalities seems to provide the proper setting in which to study questions of existence, uniqueness and regularity. However, the necessary tools for handling such problems are only now being developed and hence there have been only limited applications in elasticity.

Chapter 8

Uniqueness Theorems in Elastodynamics.
Relations with Existence, Stability, and Boundedness of Solutions

There are comparatively few results in the literature on the uniqueness of elastodynamic solutions. Nevertheless, in spite of their small number, these results provide a level of comprehension not yet attained in the static theory. Most of these contributions have been concerned exclusively with solutions that are twice continuously differentiable and have imposed, in addition to the major symmetry on the elasticities, the extra (usually non-essential) minor symmetry

$$c_{ijkl} = c_{jikl}. \tag{8.0.1}$$

The classical result, due to Neumann [1885], states that the initial-mixed boundary value problem for finite regions has a unique solution provided

$$c_{ijkl} \xi_{ij} \xi_{kl} \geqq 0 \tag{8.0.2}$$

for arbitrary tensors ξ_{ij}. Under the assumption of uniform density and elasticities, Gurtin and Toupin [1965], extending a result of Gurtin and Sternberg [1961b], proved that the displacement boundary value problem for finite regions has a unique solution provided

$$c_{ijkl} \xi_i \xi_k \eta_j \eta_l \geqq 0, \quad \text{for } \xi, \eta \neq 0. \tag{8.0.3}$$

These conclusions were all derived by means of energy arguments. A completely different approach based on properties of analytic functions was used by Hayes and Knops [1968] to show that the result of Gurtin and Toupin remains true if

$$c_{ijkl} \xi_i \xi_k \eta_j \eta_l < 0, \quad \text{for } \xi, \eta \neq 0. \tag{8.0.4}$$

Observe that condition (8.0.4) is necessary and sufficient for all plane waves to travel with purely imaginary speeds. Condition (8.0.3), implying that the equilibrium equations are semi-strongly elliptic, states that all plane waves travel with real speeds.[1] On the other hand, the classical

---
[1] See Toupin and Bernstein [1961].

condition (8.0.2) requires the elasticities to be positive-semi-definite, and makes no direct statement about the realness of wave-speeds.

Another form of energy argument, which relies on the conservation of total energy, has been used in two different ways by Brun [1965, 1969] and by Knops and Payne [1968a] to establish uniqueness in the initial-mixed boundary value problem for finite regions. This result follows, provided only that the elasticities satisfy the symmetry condition

$$c_{ijkl} = c_{klij}. \tag{8.0.5}$$

Brun dealt with the classical solution, but his method may be extended to weak solutions, this being the situation treated by Knops and Payne.

A further development, due to Wheeler and Sternberg [1968], generalised the conclusions of Neumann and of Gurtin and Sternberg to problems in which the material occupies either an exterior domain or a domain whose boundary extends to infinity.

All these results are considered in somewhat more detail in the present chapter. In Section 8.5 we treat certain non-standard problems which include those with ambiguous conditions analogous to ones considered in Section 7.3 for the static theory. By way of concluding the tract, the last section explores relations between the uniqueness of a solution and its existence, stability, and boundedness. We thus finally relate the study of uniqueness explicitly with some other fields of interest.

8.1 The Initial Displacement and Mixed-Boundary Value Problems. Energy Arguments

As outlined in Chapter 3, Neumann [1885], using standard Green's identity arguments, established the uniqueness of the classical solution to the initial displacement, traction, and mixed problems for non-homogeneous, anisotropic elastic solids occupying bounded regions, provided the elasticities are positive-semi-definite (i.e., obey 8.0.2) and satisfy the symmetries

$$c_{ijkl} = c_{jikl} = c_{klij}. \tag{8.1.1}$$

It is worth remarking here that Neumann's result remains true for those problems in which

$$\int_0^t \int_{\partial B(\eta)} n_j c_{ijkl} \frac{\partial v_k}{\partial x_l} \frac{\partial v_i}{\partial t} \, dS \, d\eta \leq 0, \tag{8.1.2}$$

where v_i is the difference between any two possible solutions of the problem. Hill [1967, Eq. (3.4)] has observed that when the *surface integral* in (8.1.2) is non-positive then certain conditional data are characterised, e.g., contact with passive constraints.

An improvement on Neumann's result in the displacement problem was obtained by Gurtin and Sternberg [1961 b]. They observed that the Cosserat form (3.9) of the potential energy is also appropriate in the initial-displacement boundary value problem for an isotropic homogeneous elastic solid (a fact originally recorded by Kelvin [1888]). They hence deduced that the pseudo-energy $\bar{E}(u)$, given by

$$\bar{E}(u) = \frac{\mu}{2} \int_B \left\{ \rho \frac{\partial u_i}{\partial t} \frac{\partial u_i}{\partial t} + \left(\frac{\partial u_i}{\partial x_j} - \frac{\partial u_j}{\partial x_i} \right) \frac{\partial u_i}{\partial x_j} + \frac{2(1-\sigma)}{1-2\sigma} \frac{\partial u_j}{\partial x_j} \frac{\partial u_k}{\partial x_k} \right\} dx, \quad (8.1.3)$$

is conserved, and were then able to derive uniqueness (of the classical solution) in bounded regions, provided

$$\mu \geq 0, \quad -\infty < \sigma < \tfrac{1}{2}, \quad 1 \leq \sigma < \infty. \quad (8.1.4)$$

Their argument is an extension of that due to Neumann. Gurtin and Toupin [1965] extend this result to a homogeneous anisotropic elastic solid, proving uniqueness of the classical solution under condition (8.0.3). Condition (8.1.4) is obtained from (8.0.3) by specialisation to isotropy. The proof adopted by Gurtin and Toupin uses conservation of energy and follows closely that of Section 4.1.2; consequently it is not reproduced here. Wheeler and Sternberg [1968] show that the first two of the above results remain valid for solids occupying what they term "regular regions". A regular region is defined as an open region whose intersection with a nested sequence of concentric balls, possessing a minimum non-zero radius, is a connected set with boundary consisting of a finite number of "closed regular surfaces" in the sense of Kellogg [1953, p. 112]. Such regions may therefore be either interior or exterior to the boundary, or have the boundary (not necessarily planar) extending to infinity. The uniqueness theorem, based upon a generalised energy identity also proved by the authors, holds for a classical solution and requires the prescription of the relevant field components on all parts of the surface of the solid, including the portion at infinity. The result of Gurtin and Toupin may be similarly extended to regular regions but is now valid only in the class of solutions possessing the required Fourier transforms.

8.2. The Initial-Displacement Boundary Value Problem. Analyticity Arguments

The energy arguments of the previous section are by no means the only method of establishing uniqueness in elastodynamical problems, and in this section we present another approach utilising the uniqueness property of analytic functions. The result was obtained by Hayes and Knops [1968] and is repeated here with some generalisation.

We discuss in detail the initial-displacement boundary value problem and then comment later on various extensions. We consider a non-homogeneous, anisotropic, elastic solid occupying a bounded region B of euclidean space. The elasticities are supposed to be analytic in B and to satisfy at every point in B the definiteness condition

$$c_{ijkl}\, \xi_i\, \xi_k\, \eta_j\, \eta_l < 0 \tag{8.2.1}$$

for all non-zero ξ_i and η_i. The elasticities are *not* required to obey any symmetry condition, so that, in particular, the following conclusions are valid for Cauchy elasticity. The density may be non-uniform, but is assumed positive and analytic in B. Uniqueness is sought in the class of solutions, continuous together with their first time derivatives in $\bar{B} \times [0, T]$ and twice continuously differentiable in the open region $B \times (0, T)$.

By virtue of the linearity of the governing equations, it clearly suffices in proving uniqueness to show that the homogeneous problem:

$$\rho\frac{\partial^2 v_i}{\partial t^2} - \frac{\partial}{\partial x_j}\left(c_{ijkl}\frac{\partial v_k}{\partial x_l}\right) = 0 \quad \text{in } B \times (0, T) \tag{8.2.2}$$

$$v_i(\mathbf{x}, t) = 0, \qquad \text{on } \partial B \times [0, T], \tag{8.2.3}$$

$$v_i(\mathbf{x}, 0) = \frac{\partial v_i}{\partial t}(\mathbf{x}, 0) = 0, \qquad \text{on } \bar{B}(0), \tag{8.2.4}$$

must possess only the zero solution. In (8.2.4), and subsequently, the symbol $B(t)$ refers to the region B at time t. Now, condition (8.2.1) is sufficient to ensure that Eq. (8.2.2) has no real characteristics, i.e., at no point on any surface γ in $B \times (0, T)$ does

$$\det |-c_{ijkl}\, n_j\, n_l + \rho\, \delta_{ik}\, n_t^2| = 0, \tag{8.2.5}$$

where n_i and n_t are the direction cosines of the unit normal on γ. Hence, by Holmgren's theorem [2] the problem has a unique solution. We emphasize again that the present argument requires no symmetry whatsoever on the elasticities.

We note that the above technique is valid and actually yields uniqueness without the prescription of data of any kind on ∂B. Thus, uniqueness holds for any standard or non-standard initial-boundary value problem. Likewise, if the displacement is prescribed on $B(0)$, and the velocity is prescribed on $B(t_1)$ (or *vice versa*), where $0 < t_1 < T$, and any standard

---

[2] See footnote on p. 36 or John [1964 p. 47].

boundary conditions are given on $\partial B \times [0, t_1]$, then uniqueness may easily be established in $B \times [0, T]$.

All the proofs for uniqueness so far discussed have appealed to some definiteness property on the elasticities. We next turn to methods requiring no such restriction.

8.3 The Initial-Mixed Boundary Value Problem for Bounded Regions. Further Arguments

We again consider a non-homogeneous, anisotropic, elastic solid, occupying a bounded region B of euclidean space with smooth surface ∂B, such that the elasticities obey the symmetry condition

$$c_{ijkl} = c_{klij} \tag{8.3.1}$$

and are further supposed to be continuous on B. The density, which may be non-uniform, is supposed to be of constant sign in $\bar{B}$. Uniqueness is here established in the class of classical solutions, defined in Section 2.2 and the previous section, but the techniques we employ can be extended to the class of weak solutions described in Section 2.2. Details for one such method were given by Knops and Payne [1968a]. We note that the class of weak solutions is large enough to include solutions to the practically important group of problems concerned with composite anisotropic elastic solids.

To establish uniqueness we adopt the device of the previous section and show that the homogeneous problem

$$\rho \frac{\partial^2 v_i}{\partial t^2} - \frac{\partial}{\partial x_j} \left(c_{ijkl} \frac{\partial v_k}{\partial x_l} \right) = 0 \quad \text{in } B \times (0, T], \tag{8.3.2}$$

$$v_i = 0, \qquad \text{on } \overline{\partial B_1} \times [0, T],$$

$$n_j c_{ijkl} \frac{\partial v_k}{\partial x_l} = 0, \qquad \text{on } \partial B_2 \times [0, T], \tag{8.3.3}$$

$$v_i(\mathbf{x}, 0) = \frac{\partial v_i}{\partial t}(\mathbf{x}, 0) = 0, \qquad \text{on } \bar{B}(0), \tag{8.3.4}$$

subject to (8.3.1), possesses only the zero solution. In (8.3.3), ∂B_1 and ∂B_2 are disjoint subsets of ∂B such that $\overline{\partial B_1} \cup \partial B_2 = \partial B$, and n_i are the components of the unit outward normal on ∂B_2. Although only boundary conditions (8.3.3) are treated, other standard types may also be considered. Some non-standard types are mentioned in Section 8.5.

We describe two methods of proof. The first, essentially due to Brun [1965, 1969], rests upon the "virtual work" theorem and derives uniqueness under the mild condition (8.3.1). The second method, due to Knops and Payne [1968a], depends upon convexity arguments which are currently of wide use in the study of improperly posed problems (see, e.g., Payne [1966] and the references cited there). Although the second approach produces no improvement upon the results established by the first method, it is nevertheless included here not only for its own intrinsic interest but also because of its later use in Section 8.5 to discuss certain non-standard problems.

It is worth remarking that extensions of both methods have been used to prove uniqueness in the corresponding linear theory of dynamic, coupled thermoelasticity. The interested reader should consult Brun [1965, 1969] and Knops and Payne [1970].

The treatment of uniqueness by the first method begins by choosing η and τ in the interval $0 < \eta < 2\tau < T$, and then multiplying Eq. (8.3.2) at time $t = \eta$ by $v_i(\mathbf{x}, 2\tau - \eta)$:

$$v_i(\mathbf{x}, 2\tau - \eta) \frac{\partial}{\partial x_j} \left\{ c_{ijkl} \frac{\partial v_k}{\partial x_l}(\mathbf{x}, \eta) \right\} = \rho \, v_i(\mathbf{x}, 2\tau - \eta) \, \ddot{v}_i(\mathbf{x}, \eta).$$

A superposed dot indicates differentiation with respect to the displayed time argument. An integration by parts and use of (8.3.3) leads immediately to

$$\int_0^\tau \int_B \left[c_{ijkl} \frac{\partial v_i}{\partial x_j}(\mathbf{x}, 2\tau - \eta) \frac{\partial v_k}{\partial x_l}(\mathbf{x}, \eta) + \rho \, v_i(\mathbf{x}, 2\tau - \eta) \, \ddot{v}_i(\mathbf{x}, \eta) \right] dx \, d\eta = 0.$$

Similarly, by evaluating (8.3.2) at time $t = 2\tau - \eta$ and multiplying by $v_i(\mathbf{x}, \eta)$, a repetition of the previous argument yields

$$\int_0^\tau \int_B \left[c_{ijkl} \frac{\partial v_i}{\partial x_j}(\mathbf{x}, \eta) \frac{\partial v_k}{\partial x_l}(\mathbf{x}, 2\tau - \eta) + \rho \, v_i(\mathbf{x}, \eta) \, \ddot{v}_i(\mathbf{x}, 2\tau - \eta) \right] dx \, d\eta = 0.$$

Because of the symmetry condition (8.3.1), the difference of these last two equations gives

$$\int_0^\tau \int_B \rho \left[v_i(\mathbf{x}, 2\tau - \eta) \, \ddot{v}_i(\mathbf{x}, \eta) - v_i(\mathbf{x}, \eta) \, \ddot{v}_i(\mathbf{x}, 2\tau - \eta) \right] dx \, d\eta = 0,$$

which, upon integrating by parts once with respect to time and using (8.3.4), reduces to

$$\int_{B(\tau)} \rho \, v_i \frac{\partial v_i}{\partial \tau} \, dx = 0.$$

A further integration and the use of (8.3.4) finally produces

$$\int_{B(t)} \rho\, v_i\, v_i\, dx = 0,$$

from which uniqueness easily follows.

To obtain uniqueness by the second method, we define the function $G(t)$ by

$$G(t) = \log F(t), \qquad F(t) = \int_{B(t)} \rho\, v_i\, v_i\, dx. \tag{8.3.5}$$

Now, as before, uniqueness is clearly implied by $F(t) \equiv 0$ for all $t \in [0, T]$. Hence, we assume that there is an interval $0 \le t_1 < t < t_2 \le T$ on which $F(t) > 0$. On this interval we show that $G(t)$ is a convex function of t, i.e., we show

$$F^2 \frac{d^2 G}{dt^2} \equiv F \frac{d^2 F}{dt^2} - \left(\frac{dF}{dt}\right)^2 \ge 0, \qquad t_1 < t < t_2. \tag{8.3.6}$$

Obviously, without loss we may take $F(t_1) = 0$. Now,

$$\frac{dF}{dt} = 2 \int_{B(t)} \rho\, v_i \frac{\partial v_i}{\partial t}\, dx \tag{8.3.7}$$

and so

$$\frac{d^2 F}{dt^2} = 2 \int_{B(t)} \left(\rho \frac{\partial v_i}{\partial t} \frac{\partial v_i}{\partial t} + \rho\, v_i \frac{\partial^2 v_i}{\partial t^2} \right) dx. \tag{8.3.8}$$

In the second term on the right, we may substitute for the acceleration from Eq. (8.3.2) and integrate by parts to obtain

$$\frac{d^2 F}{dt^2} = 2 \int_{B(t)} \left(\rho \frac{\partial v_i}{\partial t} \frac{\partial v_i}{\partial t} - c_{ijkl} \frac{\partial v_i}{\partial x_j} \frac{\partial v_k}{\partial x_l} \right) dx. \tag{8.3.9}$$

Here we have used the boundary conditions (8.3.3). On combining (8.3.5), (8.3.7) and (8.3.9) we find

$$F \frac{d^2 F}{dt^2} - \left(\frac{dF}{dt}\right)^2 = 4 \left\{ \int_B \rho\, v_i\, v_i\, dx \int_B \rho \frac{\partial v_i}{\partial t} \frac{\partial v_i}{\partial t}\, dx - \left[\int_B \rho\, v_i \frac{\partial v_i}{\partial t}\, dx \right]^2 \right\}$$

$$- 2 \int_B \rho\, v_i\, v_i\, dx \int_B \left(\rho \frac{\partial v_i}{\partial t} \frac{\partial v_i}{\partial t} + c_{ijkl} \frac{\partial v_i}{\partial x_j} \frac{\partial v_k}{\partial x_l} \right) dx. \tag{8.3.10}$$

In (8.3.10), the last term vanishes, since by conservation of energy and the initial conditions (8.3.4) the total energy of the system is zero. The first term on the right of (8.3.10) is non-negative by virtue of Schwarz's inequality. Thus, inequality (8.3.6) is established. Together with the continuity of $F(t)$ this implies, after integration,

$$F(t) \le [F(t_2)]^{\frac{t - t_1}{t_2 - t_1}} [F(t_1)]^{\frac{t_2 - t}{t_2 - t_1}}, \qquad t_1 \le t \le t_2. \tag{8.3.11}$$

But it has been assumed already that $F(t_1)=0$, so that (8.3.11) immediately shows that $F(t)=0$ for $t_1 \leq t \leq t_2$, contrary to hypothesis. Hence, $F(t)\equiv 0$ for all $t<T$, and by continuity we see that $F(T)\equiv 0$. Finally, since v_i is assumed continuous in $\bar{B} \times [0, T]$, we may conclude that $v_i \equiv 0$ in $\bar{B} \times [0, T]$, and uniqueness is proved.

That the symmetry condition (8.3.1) is not necessary for uniqueness of the above problems may be seen from the following examples, where uniqueness continues to hold but condition (8.3.1) is violated. In the first two examples the elasticities are taken to be uniform.

a) We examine the displacement problem in a bounded region and let

$$c_{ijkl}=\delta_{ik}\,\delta_{j1}\,\delta_{12}, \tag{8.3.12}$$

so that under zero body force the governing equations become

$$\frac{\partial^2 v_i}{\partial x_1 \, \partial x_2}=\rho\,\frac{\partial^2 v_i}{\partial t^2}. \tag{8.3.13}$$

Uniqueness may be easily established by applying to (8.3.13) the convexity argument previously used.

b) In this example we again consider a bounded region, but with boundary conditions of mixed type. If we take uniform elasticities satisfying

$$c_{ijkl}=-c_{ilkj},$$

the governing equations under zero body force become

$$\frac{\partial^2 v_i}{\partial t^2}=0, \tag{8.3.14}$$

and uniqueness follows trivially. In fact, boundary conditions are not essential for uniqueness.

In the above two examples it is worth noting that uniqueness is established even though the elasticities neither possess major symmetry (8.3.1) nor satisfy a definiteness condition. Our third example[3] shows by means of transformation of axes, that uniqueness and conservation of total energy are sometimes compatible with violation of condition (8.3.1).

c) We refer the elastic body to an oblique set of axes, thus obtaining for the components u^i of the displacement the equations

$$\frac{\partial}{\partial x_j}\left(c^{ij}{}_k{}^l\,\frac{\partial u^k}{\partial x_l}\right)=\rho\,\frac{\partial^2 u^i}{\partial t^2},$$

---

[3] We are grateful to Professor J. L. Ericksen for suggesting this example.

where the body force is taken to be zero. By choosing in (8.3.5) the function $F(t)$ to be

$$F(t) = \int_B \rho\, g_{ij}\, v^i\, v^j\, dx,$$

where g_{ij} are the constant, symmetric components of the metric tensor, it may easily be seen that the convexity argument continues to be valid provided

$$g_{pi}\, c^{ij}{}_k{}^l = g_{ki}\, c^{il}{}_p{}^j. \tag{8.3.15}$$

Condition (8.3.15) implies conservation of total energy, and is satisfied, for instance, by

$$c^{ij}{}_k{}^l = g^{ij}\, \delta_k^l$$

where

$$g_{pi}\, g^{ij} = \delta_p^j.$$

We remark finally that a similar example may be constructed in which the axes are rotated.[4]

8.4 Summary of Existing Results in the Uniqueness of Elastodynamic Solutions

The results of the last three sections may be briefly summarised. All the standard initial-boundary value problems for non-homogeneous anisotropic elastic solids occupying bounded regions have a unique classical solution provided

$$c_{ijkl} = c_{klij}. \tag{8.4.1}$$

If this condition is violated, then uniqueness holds provided ρ and the elasticities are analytic and

$$c_{ijkl}\, \xi_i\, \xi_k\, \eta_j\, \eta_l < 0 \tag{8.4.2}$$

for non-zero ξ_i, η_i. Neither condition is necessary for uniqueness. In unbounded regions, or in regions whose boundary extends to infinity, there is uniqueness of the classical solution in the standard problem, provided the elasticities satisfy either (8.4.2) or the positive-semi-definiteness condition

$$c_{ijkl}\, \xi_{ij}\, \xi_{kl} \geq 0, \tag{8.4.3}$$

for non-zero tensors ξ_{ij}.

For the displacement problem in these regions, the classical solution is unique, provided either (8.4.2) holds or, in the isotropic case, if

$$\mu \geq 0, \quad -\infty < \sigma < \tfrac{1}{2}, \quad 1 \leq \sigma < \infty. \tag{8.4.4}$$

---

[4] This was indicated to us by Dr. M. Hayes.

The condition (8.4.4) may be extended to homogeneous anisotropic media by combining the methods of Gurtin and Toupin and of Wheeler and Sternberg. It then becomes

$$c_{ijkl}\,\xi_i\,\xi_k\,\eta_j\,\eta_l \geqq 0$$

for non-zero ξ_i, η_i, but this condition is sufficient for the uniqueness only of solutions possessing the required Fourier transforms.

8.5 Non-Standard Problems, including those with Ambiguous Conditions

Certain non-standard initial-boundary value problems may also be treated by convexity arguments. Two such groups of problems are considered, the first having non-standard data prescribed on the boundary, the second having ambiguous conditions on part of the boundary with standard conditions on the remainder. To be consistent, we include in the second group problems involving solutions of certain variational inequalities, and thus this group forms the dynamic counterpart of the Signorini problems discussed in Section 7.3. However, the uniqueness proof for the solution to the variational inequality must be based on energy arguments.

We begin by establishing uniqueness in the first group. To do this, we use (8.3.8) to replace (8.3.10) by

$$F\frac{d^2 F}{dt^2} - \left(\frac{dF}{dt}\right)^2 \geqq -2F \int_{B(t)} \left(\rho\frac{\partial v_i}{\partial t}\frac{\partial v_i}{\partial t} - \rho\,v_i\frac{\partial^2 v_i}{\partial t^2}\right) dx. \quad (8.5.1)$$

Thus the previous argument remains valid for any conditions which imply that

$$H(t) \equiv \int_{B(t)} \left(\rho\frac{\partial v_i}{\partial t}\frac{\partial v_i}{\partial t} - \rho\,v_i\frac{\partial^2 v_i}{\partial t^2}\right) dx \leqq 0. \quad (8.5.2)$$

Now, a time differentiation of (8.5.2), followed by substitution from (8.3.2) and an integration by parts, yields

$$\frac{dH}{dt} = \int_{\partial B(t)} \left(n_j c_{ijkl}\frac{\partial v_k}{\partial x_l}\frac{\partial v_i}{\partial t} - n_j c_{ijkl}\,v_i\frac{\partial^2 v_k}{\partial x_l\,\partial t}\right) dS. \quad (8.5.3)$$

Thus, for instance, if either

a) $n_j c_{ijkl}\dfrac{\partial v_k}{\partial x_l} = \alpha(\mathbf{x}, t)\,v_i$ on ∂B, where $\alpha(\mathbf{x}, t)$ is a scalar function of position and time satisfying $\partial\alpha/\partial t \geqq 0$ on ∂B, or

b) $n_j c_{ijkl} \dfrac{\partial v_k}{\partial x_l} = A_{ij}(\mathbf{x}, t) v_j$ *on* ∂B, *where* $A_{ij}(\mathbf{x}, t)$ *is a symmetric tensor function of position and time with* $\partial A_{ij}/\partial t$ *positive-definite on* ∂B,

it follows that $dH/dt \leq 0$ and therefore that

$$H(t) \leq H(0).$$

But by the continuity of v_i and the initial conditions (8.3.4) we have that $H(0) \equiv 0$, and so (8.5.2) is proved. The boundary conditions introduced in (a) and (b) correspond to those in generalised elastic support problems.

In those problems of the second group having ambiguous boundary conditions we can sometimes establish uniqueness by again demonstrating that $H(t)$ is non-positive. The method is illustrated by two different examples in each of which Σ denotes a portion of the boundary ∂B. As before, no definiteness assumptions are made on the elasticities, but in the case when these are positive-semi-definite, uniqueness can be established in the first example by an energy argument requiring slightly fewer boundary conditions.

Thus, we begin by letting the function u_i, of class

$$K_1: \{C^2(B \times (0, T) \cup C^0(\bar{B} \times [0, T]) \cup H^2(B \times (0, T))\},$$

be a solution of the system

$$\rho \frac{\partial^2 u_i}{\partial t^2} - \frac{\partial}{\partial x_j}\left(c_{ijkl}\frac{\partial u_k}{\partial x_l}\right) = \rho F_i \quad \text{in } B \times (0, T]$$

$$\sigma_{ij} n_j n_k - \sigma_{kj} n_j n_i = f_{ik} \quad \text{on } \partial B \times (0, T]$$

$$\sigma_{ij} n_i n_j = g \quad \text{on } \partial B/\Sigma \times (0, T] \tag{8.5.4}$$

$$u_i(\mathbf{x}, 0) = f_i, \quad \frac{\partial u_i}{\partial t}(\mathbf{x}, 0) = g_i \quad \text{on } B(0)$$

and the following two sets of ambiguous conditions: at every point on $\bar{\Sigma}$ either

$$\frac{\partial u_i}{\partial t} n_i > f, \quad \sigma_{ij} n_i n_j = 0 \tag{8.5.5a}$$

or

$$\frac{\partial u_i}{\partial t} n_i = f, \quad \sigma_{ij} n_i n_j \geq 0 \tag{8.5.5b}$$

and either

$$u_i n_i < h, \quad \sigma_{ij} n_i n_j = 0 \tag{8.5.6a}$$

or

$$u_i n_i = h, \quad \sigma_{ij} n_i n_j \geq 0. \tag{8.5.6b}$$

Here, the data consist of the assigned functions $F_i(\mathbf{x}, t)$, $f_{ik}(\mathbf{x}, t)$, $g(\mathbf{x}, t)$, $f_i(\mathbf{x})$, $g_i(\mathbf{x})$, $f(\mathbf{x}, t)$ and $h(\mathbf{x}, t)$. Of course, no solution can exist without strong compatibility conditions on the data. The stress σ_{ij} is related to u_i through (2.1.10), i.e.,

$$\sigma_{ij} = c_{ijkl} \frac{\partial u_k}{\partial x_l}, \tag{8.5.7}$$

and clearly we must take (8.5.5a) with (8.5.6a), (8.5.5b) with (8.5.6b).

With the understanding that the elasticities satisfy appropriate continuity and differentiability properties, we now prove the following theorem.

Theorem 8.5.1. *Provided the elasticities satisfy the major symmetry $c_{ijkl} = c_{klij}$, the system (8.5.4), (8.5.5) and (8.5.6) has at most one solution of class K_1.*

We assume the existence of two solutions u_i^1 and u_i^2 and letting $v_i = u_i^1 - u_i^2$ observe that v_i satisfies (8.5.4) with homogeneous data. Furthermore, on Σ, we have for σ_{ij} defined in terms of v_i according to (8.5.7)

$$\sigma_{ij} n_j \frac{\partial v_i}{\partial t} \equiv \sigma_{ij} n_i n_j n_k \frac{\partial v_k}{\partial t}, \tag{8.5.8}$$

$$\sigma_{ij} n_j v_i \equiv \sigma_{ij} n_i n_j n_k v_k. \tag{8.5.9}$$

On the other hand, at an arbitrary point P of Σ we may assume without loss that $\dfrac{\partial u_i^1}{\partial t} n_i \geq \dfrac{\partial u_i^2}{\partial t} n_i$. Then one of three possibilities arises. Either

$$\text{(i)} \quad \frac{\partial u_i^1}{\partial t} n_i > f, \quad \frac{\partial u_i^2}{\partial t} n_i > f,$$

or

$$\text{(ii)} \quad \frac{\partial u_i^1}{\partial t} n_i > f, \quad \frac{\partial u_i^2}{\partial t} n_i = f,$$

or

$$\text{(iii)} \quad \frac{\partial u_i^1}{\partial t} n_i = f, \quad \frac{\partial u_i^2}{\partial t} n_i = f.$$

We separately examine consequences of these conditions in the light of (8.5.5) and (8.5.6).

First, in case (i), from (8.5.5a) we find that

$$\sigma_{ij} n_i n_j = 0.$$

Secondly, in case (ii), it follows from (8.5.5a) that the previous identity holds, or alternatively, from (8.5.5a) and (8.5.5b), that we have

$$\sigma_{ij} n_i n_j < 0.$$

In this event, we gather from (8.5.6a) and (8.5.6b) respectively that

$$u_i^1 n_i \leqq h, \quad u_i^2 n_i = h,$$

and therefore

$$v_i n_i \leqq 0.$$

In case (iii), four situations are possible, corresponding to the manner in which (8.5.5b) is satisfied. Thus, we may deduce from (8.5.5b) that *either* (a)

$$\sigma_{ij}^1 n_i n_j > 0, \quad \sigma_{ij}^2 n_i n_j > 0,$$

where σ_{ij}^1, σ_{ij}^2 are the stress components associated with u_i^1, u_i^2, respectively, from which it follows by (8.5.6b) that

$$u_i^1 n_i = u_i^2 n_i = h,$$

and therefore

$$v_i n_i = 0,$$

or (b),

$$\sigma_{ij}^1 n_i n_j > 0, \quad \sigma_{ij}^2 n_i n_j = 0$$

which from (8.5.6b) shows that

$$u_i^1 n_i = h, \quad u_i^2 n_i < h,$$

and hence

$$v_i n_i > 0;$$

or (c),

$$\sigma_{ij}^1 n_i n_j = 0, \quad \sigma_{ij}^2 n_i n_j > 0,$$

which, similarly to (b), yields

$$v_i n_i < 0;$$

or (d),

$$\sigma_{ij}^1 n_i n_j = \sigma_{ij}^2 n_i n_j = 0.$$

These conclusions taken in conjunction with (8.5.8) and (8.5.9) then demonstrate that

$$\sigma_{ij} n_j \frac{\partial v_i}{\partial t} \leqq 0 \quad \text{on } \Sigma, \tag{8.5.10}$$

and

$$\sigma_{ij} n_j v_i \geqq 0 \quad \text{on } \Sigma. \tag{8.5.11}$$

We have now assembled all the preliminaries needed in the proof of the theorem, and we proceed as follows. Introduce the energy expression

$$E(t) = \frac{1}{2} \int_{B(t)} \left(\rho \frac{\partial v_i}{\partial t} \frac{\partial v_i}{\partial t} + c_{ijkl} \frac{\partial v_i}{\partial x_j} \frac{\partial v_k}{\partial x_l} \right) dx \tag{8.5.12}$$

and with the help of (8.5.4) write (8.5.2) in the form

$$H(t) = 2E(t) - \int\limits_{\partial B(t)} n_j \sigma_{ij} v_i \, dS$$

$$= 2E(t) - \int\limits_{\Sigma(t)} n_j \sigma_{ij} v_i \, dS. \tag{8.5.13}$$

Immediately from (8.5.11) we obtain

$$H(t) \leq 2E(t). \tag{8.5.14}$$

Next, a time differentiation of (8.5.12) together with (8.5.10) yields

$$\frac{dE(t)}{dt} = \int\limits_{\partial B(t)} \sigma_{ij} n_j \frac{\partial v_i}{\partial t} \, dS = \int\limits_{\Sigma(t)} \sigma_{ij} n_j \frac{\partial v_i}{\partial t} \, dS \leq 0, \tag{8.5.15}$$

and therefore

$$E(t) \leq E(0) \equiv 0, \tag{8.5.16}$$

the second equality holding by virtue of the homogeneous initial data. Thus, finally, from (8.5.14) and (8.5.16) we obtain

$$H(t) \leq 0$$

and hence (8.5.2) is proved.

When the elasticities are assumed to be positive-semi-definite uniqueness may be established without imposing the ambiguous conditions (8.5.6). For clearly, (8.5.16) is still valid but now by hypothesis, $E(t) \geq 0$, and so we conclude that

$$E(t) \equiv 0 \qquad 0 \leq t \leq T \tag{8.5.17}$$

which implies $v_i \equiv 0$ in $\bar{B} \times [0, T]$. We have therefore proved the following

Theorem 8.5.2. *Provided the elasticities satisfy the major symmetry* $c_{ijkl} = c_{klij}$ *and are also positive-semi-definite,[5] the system (8.5.4) and (8.5.5) has at most one solution of class* K_1.

We now consider an example of a problem characterized as the solution of a variational inequality. Let us define K_2 as the set

$$K_2 = \left\{ \varphi_i \in H_0^2(B \times (0, T)); \, \frac{\partial \varphi_i}{\partial t} \, \frac{\partial \varphi_i}{\partial t} \leq k \right\}, \tag{8.5.18}$$

for positive constant k. Then we seek a solution $u_i \in K_2$ of the variational inequality

$$\int\limits_0^t \int\limits_{B(\eta)} \left(\frac{\partial^2 u_i}{\partial \eta^2} - \frac{\partial}{\partial x_j} \left(c_{ijkl} \frac{\partial u_k}{\partial x_l} \right) - \rho F_i \right) \left(\frac{\partial \varphi_i}{\partial \eta} - \frac{\partial u_i}{\partial \eta} \right) dx \, d\eta \geq 0 \tag{8.5.19}$$

---

[5] In addition, we assume that appropriate continuity and differentiability conditions are satisfied.

for all $\varphi_i \in K_2$, subject to the prescribed initial conditions

$$u_i(\mathbf{x}, 0) = f_i(\mathbf{x}), \qquad \frac{\partial u_i}{\partial t}(\mathbf{x}, 0) = g_i(\mathbf{x}). \tag{8.5.20}$$

We consider only the question of uniqueness of solution and establish the following theorem:

Theorem 8.5.3. *Provided the elasticities are positive semi-definite and satisfy the major symmetry* $c_{ijkl} = c_{klij}$, *then problem (8.5.19) and (8.5.20) has at most one solution of class* K_2.

To prove this theorem assume that two solutions u_i^1 and u_i^2 exist and write the inequality (8.5.19) for $u_i = u_i^1$ taking $\varphi_i = u_i^2$, and then for $u_i = u_i^2$ taking $\varphi_i = u_i^1$. Addition of these two inequalities leads to the inequality for $v_i = u_i^1 - u_i^2$:

$$\int_0^t \int_{B(\eta)} \left[\frac{\partial^2 v_i}{\partial \eta^2} - \frac{\partial}{\partial x_j} \left(c_{ijkl} \frac{\partial v_k}{\partial x_l} \right) \right] \frac{\partial v_i}{\partial \eta} \, dx \, d\eta \leq 0, \tag{8.5.21}$$

which obviously yields

$$E(t) \leq E(0) \equiv 0. \tag{8.5.22}$$

But by hypothesis, $E(t) \geq 0$, and so

$$E(t) \equiv 0, \qquad 0 \leq t \leq T,$$

which guarantees the identical vanishing of v_i in $\bar{B} \times [0, T]$. The theorem is thus proved.

Quite clearly, the problems we have indicated are of limited physical significance, but it appears likely that a number of physically interesting dynamical problems of ambiguous character could be handled by the methods indicated here. Again it should be remarked that the study of such ambiguous problems has just recently been initiated and in fact most, if not all, of the investigations in elasticity have been limited to equilibrium solutions.

8.6 Stability, Boundedness, Existence and Uniqueness

In this final section, we collect together some results in the literature connecting the concepts of stability, boundedness, existence and uniqueness. Until now, this Tract has examined exclusively the uniqueness question. A considerable literature testifies to the interest aroused in the topic of the existence of a solution to the linearised problem; see e.g. Korn [1908], Friedrichs [1947], Browder [1954] and Fichera [1950, 1965]. There is also an extensive literature on the stability of the solution to a dynamical system although, at least in elasticity theory, there has not always been agreement on the definition of stability. In this respect we mention in particular the work of Hill [1957], Holden [1964], Koiter [1965], Shield [1965], Movchan [1960, 1963], Slobokin [1962], Zorski [1962], Knops and Wilkes [1966], Gilbert and Knops [1967], and Knops and Payne [1968 b].

We first examine relations between the three notions of stability, boundedness and uniqueness, and follow the treatment presented by Gilbert and Knops [1967]. Their investigation, based upon the definition of stability due to Liapounov [1892] as generalised to continuous systems by Movchan [1960], is carried out in abstract terms, but for our purposes it is sufficient to consider the following: Let $X(t)$ denote the space of linear vector functions defined in a spatial region B over a time interval $\mathcal{T}$ which includes the origin. The functions are assumed twice continuously differentiable in the spatial and time variables. Elements of $X(t)$ correspond to the solutions of the differential equations which in the present case are those governing linearised elastodynamics. Let the initial data belong to the set $X(0)$, and introduce the norms $\|(\cdot)\|_0$, $\|(\cdot)\|_t$ defined on $X(0)$ and $X(t)$ respectively. The norm $\|(\cdot)\|$ given by

$$\|(\cdot)\| = \sup_{t \in \mathcal{T}} \|(\cdot)\|_t, \tag{8.6.1}$$

is adopted as the measure of the subsequent perturbations of the motion of the system, while the norm $\|(\cdot)\|_0$ is adopted as the measure of the initial perturbations. Let $u_i(x, t) \in X(t)$ be the solution whose stability is being investigated, and let $w_i(\mathbf{x}, t) \in X(t)$ be any other solution of the differential equations, (subject in our case to the same boundary conditions as u_i). Then we say that the solution u_i is

(a) *stable* if for any $\varepsilon > 0$ there exists $\delta(\varepsilon) > 0$ such that

$$\|w - u\|_0 < \delta \tag{8.6.2}$$

implies

$$\|w - u\| < \varepsilon, \tag{8.6.3}$$

(b) *asymptotically stable* if inequalities (8.6.2) and (8.6.3) hold and in addition

$$\lim_{t \to \infty} \|w - u\|_t = 0. \tag{8.6.4}$$

As might be expected, it follows from the definition that a *stable solution is also unique*. The proof is trivial and so is omitted. (See Movchan [1960], Gilbert and Knops [1967].) A consequence of this statement is that non-uniqueness of a solution automatically implies its instability.

Gilbert and Knops [1967] also prove that the boundedness of a solution to a linear system is necessary and sufficient to insure that the solution be stable. The sufficiency of this condition is obvious. To prove necessity, note that boundedness may be alternatively defined as follows. A solution is bounded if there exist constants A and B (possibly depending on the solution) such that

$$\|w - u\|_0 \leq A \tag{8.6.5}$$

implies

$$\|w - u\| \leq B. \tag{8.6.6}$$

Now, for $\varepsilon > 0$ define the constant α by

$$\alpha = \varepsilon\, B^{-1} \tag{8.6.7}$$

and set

$$v_i^{(1)} = \alpha\, u_i, \qquad v_i^{(2)} = \alpha\, w_i. \tag{8.6.8}$$

Clearly, $v_i^{(k)}$, $k = 1, 2$, is a solution of the same system, but with initial data different from that of u_i, w_i. Substitution of (8.6.8), (8.6.7) into (8.6.5) and (8.6.6) then shows that

$$\|v^{(2)} - v^{(1)}\|_0 \leq \varepsilon\, \frac{A}{B} \equiv \delta \tag{8.6.9}$$

implies

$$\|v^{(2)} - v^{(1)}\| \leq \varepsilon, \tag{8.6.10}$$

which is just the definition of stability. The proof is therefore complete.

In the special case of linear elasticity, a relationship exists between uniqueness in the static problem and stability (and boundedness) of the solution in the dynamic problem. This can most easily be seen by noting that if $v_i(\mathbf{x})$ is a non-zero solution of the static problem with homogeneous boundary conditions, then $u_i(\mathbf{x}, t) = (1 + t)\, v_i(\mathbf{x})$ is a time-increasing solution of the corresponding dynamical problem also with homogeneous boundary conditions. Thus, non-uniqueness in static problems implies asymptotic instability (and unboundedness) of solutions to dynamic problems with the same boundary conditions. Conversely, stability of solutions in elastodynamic problems implies the uniqueness of solutions to the corresponding elastostatic problem.[6]

We next turn our attention to the connexion between existence and uniqueness, and we demonstrate that in both linear elastostatics and linear elastodynamics the existence of a classical solution implies uniqueness in a dual problem. The converse of this statement, that non-uniqueness (properly defined) implies non-existence, was established by Ericksen [1963, 1964, 1965]. The result is, of course, just a Fredholm alternative and follows from the dual problem concept discussed by e.g., Fichera [1955] and Bramble and Payne [1963 b, c].

---

[6] These remarks also apply to any linear system involving the acceleration or higher order time derivatives of the displacement.

We prove one theorem each for classical solutions in elastostatics and elastodynamics, beginning with the static case. The notation is that previously adopted.

Theorem 8.6.1. *Suppose that for arbitrary $f_i(\mathbf{x})$ there exists a solution $v_i(\mathbf{x})$ to the problem*

$$\frac{\partial}{\partial x_l}\left(c_{ijkl}\frac{\partial v_i}{\partial x_j}\right)=f_k \quad \text{in } B,$$

$$v_i=0 \quad \text{on } \overline{\partial B_1}, \tag{8.6.11}$$

$$n_l\,c_{ijkl}\frac{\partial v_i}{\partial x_j}=0 \quad \text{on } \partial B_2.$$

(If $\partial B_1=\emptyset$, certain compatibility conditions are needed to fix rigid body displacements.) Then, for specified functions $F_i(\mathbf{x})$, $g_i(\mathbf{x})$ and $h_i(\mathbf{x})$ the dual problem,

$$\frac{\partial}{\partial x_j}\left(c_{ijkl}\frac{\partial u_k}{\partial x_l}\right)=F_i \quad \text{in } B,$$

$$u_i=g_i \quad \text{on } \overline{\partial B_1}, \tag{8.6.12}$$

$$n_j\,c_{ijkl}\frac{\partial u_k}{\partial x_l}=h_i \quad \text{on } \partial B_2,$$

possesses a unique solution $u_i(\mathbf{x})$. In (8.6.11), (8.6.12), $\partial B=\overline{\partial B_1}\cup\partial B_2$.

The proof of this theorem is straightforward. Assume two solutions $u_i^{(1)}$, $u_i^{(2)}$ of (8.6.12) and put

$$w_i=u_i^{(1)}-u_i^{(2)}. \tag{8.6.13}$$

Clearly, w_i satisfies (8.6.12) with F_i, g_i and h_i all identically zero. The theorem is proved if it can be shown that existence of the solution to (8.6.11) implies the identical vanishing of w_i in B. Let us take $f_i\equiv w_i$ in B. It then follows that

$$\int_B w_k w_k\,dx=\int_B \frac{\partial}{\partial x_l}\left(c_{ijkl}\frac{\partial v_i}{\partial x_j}\right)w_k\,dx$$

$$=\oint_{\partial B} c_{ijkl}\left(\frac{\partial v_i}{\partial x_j}n_l w_k-\frac{\partial w_k}{\partial x_l}n_j v_i\right)dS \tag{8.6.14}$$

$$+\int_B v_i\frac{\partial}{\partial x_j}\left(c_{ijkl}\frac{\partial w_k}{\partial x_l}\right)dx.$$

But in view of conditions satisfied by w_i and those imposed on v_i all integrals on the right vanish. Thus, (8.6.14) gives

$$\int_B w_k w_k\,dx=0, \tag{8.6.15}$$

which implies $w_i\equiv 0$ in B, and the theorem is established.

The corollary of this theorem is obvious. Suppose a non-trivial solution w_i exists to (8.6.12) with homogeneous data. Then the left side of (8.6.14) cannot be zero, in turn implying the non-existence of a solution to (8.6.11) for $f_i \equiv w_i$. This is the result obtained by Ericksen[7] [1963, 1965] (with proper modifications for the case $\partial B_1 = \emptyset$) using the Betti reciprocal theorem. In our derivation the reciprocal theorem is not required, but we note that in those situations where this theorem is valid, i.e., when the elasticities obey the symmetry

$$c_{ijkl} = c_{klij}, \tag{8.6.16}$$

the domain and boundary operators in (8.6.11) coincide with the corresponding operators in (8.6.12).

Analogous theorems are easily proved for the elasto-dynamic system. A typical one is

Theorem 8.6.2. *Suppose that for arbitrary $f_i(\mathbf{x})$ and T there exists a solution $v_i(\mathbf{x}, t)$ to the problem*

$$\rho \frac{\partial^2 v_k}{\partial t^2} - \frac{\partial}{\partial x_l} \left(c_{ijkl} \frac{\partial v_i}{\partial x_j} \right) = 0 \quad \text{in } B \times (0, T),$$

$$v_i = 0 \quad \text{on } \overline{\partial B_1} \times [0, T],$$

$$n_l c_{ijkl} \frac{\partial v_i}{\partial x_j} = 0 \quad \text{on } \partial B_2 \times [0, T], \tag{8.6.17}$$

$$v_i(\mathbf{x}, T) = 0 \quad \text{in } B(T),$$

$$\frac{\partial v_i}{\partial t}(\mathbf{x}, T) = f_i \quad \text{in } B(T).$$

Then, for specified functions $F_i(\mathbf{x}, t)$, $g_i(\mathbf{x}, t)$, $h_i(\mathbf{x}, t)$, $\varphi_i(\mathbf{x})$ and $\psi_i(\mathbf{x})$, the dual problem

$$\rho \frac{\partial^2 u_i}{\partial t^2} - \frac{\partial}{\partial x_j} \left(c_{ijkl} \frac{\partial u_k}{\partial x_l} \right) + \rho F_i = 0 \quad \text{in } B \times (0, T),$$

$$u_i = g_i \quad \text{on } \overline{\partial B_1} \times [0, T],$$

$$n_j c_{ijkl} \frac{\partial u_k}{\partial x_l} = h_i \quad \text{on } \partial B_2 \times [0, T], \tag{8.6.18}$$

$$u_i(\mathbf{x}, 0) = \varphi_i(\mathbf{x}), \qquad \frac{\partial u_i}{\partial t}(\mathbf{x}, 0) = \psi_i(\mathbf{x}) \quad \text{on } B(0),$$

possesses a unique solution $u_i(\mathbf{x}, t)$. The density $\rho(\mathbf{x})$ is assumed positive in B.

---

[7] The result for the mixed problem is obtained by Gurtin [1971].

The proof again is straightforward. Assume two solutions $u_i^{(1)}$, $u_i^{(2)}$ of (8.6.18) and put

$$w_i = u_i^{(1)} - u_i^{(2)} \quad \text{in } B \times (0, T).$$

Now take $f_i \equiv w_i$ in (8.6.17). Then, integration by parts with respect to both space and time coordinates shows that

$$
\begin{aligned}
\int_{B(T)} \rho\, w_i w_i \, dx = {} & \int_{B(0)} \rho \left(w_i \frac{\partial v_i}{\partial t} - v_i \frac{\partial w_i}{\partial t} \right) dx + \int_{B(T)} \rho\, v_i \frac{\partial w_i}{\partial t} \, dx \\
& + \int_0^T \int_{\partial B(\eta)} c_{ijkl} \left(w_k n_l \frac{\partial v_i}{\partial x_j} - v_i n_j \frac{\partial w_k}{\partial x_l} \right) dS \, d\eta.
\end{aligned}
\tag{8.6.19}
$$

But in view of the assumed conditions on w_i and v_i all the integrals on the right of (8.6.19) vanish. Hence, the left side is zero, and we may conclude that $w_i \equiv 0$ in $B(T)$, thus establishing the theorem.

Again, non-uniqueness of the solution to (8.6.18) implies the non-existence of the solution to the dual problem, a result originally derived by Ericksen [1964] using the Betti reciprocal theorem. As before, the reciprocal theorem is not required in the present proof, but when it is valid, i.e., when (8.6.16) holds, the dual problem (8.6.17) essentially coincides with problem (8.6.18), apart from a time shift and a reversal of the time direction.

In this section we have discussed only classical solutions. Evidently, similar methods lead to alternate theorems for various weak solutions with data in appropriate spaces.

References

(Decimal numbers in parentheses following a reference give the sections in which the work is cited. Apart from 1 and 3, whole numbers indicate that the work is cited in the introductions to chapters. The numbers 1 and 3 refer to the respective entire chapter.)

1839   Green, G.: On the laws of the reflexion and rarefaction of light at the common surface of two non-crystallised media. (1837) Trans. Cambridge Phil. Soc. **7** (1839), 1–24 = Papers, 245–269. (3).

1846   Dirichlet, G. Lejeune: Über die Stabilität des Gleichgewichts. J. Reine Angew. Math. **32**, 85–88. (3).

1850   Kirchhoff, G.: Über das Gleichgewicht und die Bewegung einer elastischen Scheibe. J. Reine Angew. Math. **40**, 51–58 = Ges. Abh. (1882), 237–279. Leipzig: Barth. (3).

1855   Saint-Venant, A.-J.-C. B. de: Mémoire sur la torsion des prismes … Mem. Divers Savants Acad. Sci. Paris **14**, 233–560. (3).

1859   Kirchhoff, G.: Über das Gleichgewicht und die Bewegung eines unendlich dünnen elastischen Stabes. J. Reine Angew. Math. **56**, 285–313 = Ges. Abh. (1882), 285–316. Leipzig: Barth. (3).

1862   Clebsch, A.: Theorie der Elasticität der festen Körper. Leipzig. (3).

1873   Borchardt, C. W.: Über die Transformation der Elasticitätsgleichungen in allgemeine orthogonale Koordinaten. J. Reine Angew. Math. **76**, 45–58. (3).

1877   Kirchhoff, G.: Vorlesungen über Mathematische Physik. Mechanik. Leipzig: Teubner. (3).

1879   Thomson, Sir W., (Lord Kelvin), Tait, P.G.: Treatise on Natural Philosophy. Cambridge. Reprinted at New York: Dover Publications 1962 as Principles of Mechanics. (3).

1883   Clebsch, A.: Théorie de l'élasticité des Corps Solides. (Translated by B. de St. Venant and Flamant.) Paris: Dunod. (3).

1885   Neumann, F.: Vorlesungen über die Theorie der Elasticität der festen Körper und des Lichtäthers. Leipzig: Teubner. (See § 61) (3, 8, 8.1).

1888   Thomson, Sir W., (Lord Kelvin): Reflexion and refraction of light. Phil. Mag. (5th ser.) **26**, 414–425. (3, 8.1).

1892   Liapounov, A. M.: Problème général de la stabilité du mouvement. (Société Mathématique de Kharkow.) Translated by Davaux, E., Ann. Fac. Sci. Univ. Toulouse (2nd ser.) **9**. Reprinted at Princeton: Princeton University Press 1949. (8.6).

——   Love, A. E. H.: A Treatise on the Mathematical Theory of Elasticity, 1st edn. Cambridge: Cambridge University Press. (3).

1893   Todhunter, I., Pearson, K.: A History of the Theory of Elasticity and Strength of materials, vol. II. Cambridge. Reprinted at New York: Dover Publications 1960. (3).

1894   Poincaré, H.: Sur les équations de la physique mathématique. Rend. Cir. Mat. Palermo **8**, 57–156. (3).

1895   Chree, C.: The equilibrium of an isotropic elastic solid ellipsoid under the action
          of normal surface forces of the second degree and bodily forces derived from a
          potential of the second degree. Quart. J. Pure Appl. Math. **27**, 338–353. (3).

1898a.   Cosserat, E. and F.: Sur les équations de la théorie de l'élasticité. C.R. Acad.
          Sci. Paris **126**, 1089–1091. (3, 4.1.6).

——— b.   — Sur les fonctions potential de la theorie de l'élasticité. C.R. Acad. Sci. Paris
          126, 1129–1132. (3, 4.1.6).

——— c.   — Sur la déformation infiniment petite d'une ellipsoid élastique. C.R. Acad.
          Sci. Paris **127**, 315–318. (3, 5.1).

1901 a.   — Sur la solution des équations de l'élasticité dans le cas où les valuers des
          inconnues a la frontière sont données. C.R. Acad. Sci. Paris **133**, 145–147. (3).

——— b.   — Sur la déformation infiniment petite d'un corps élastique soumis à des forces
          donnés. C.R. Acad. Sci. Paris **133**, 271–273. Correction in ibid. **133**, 400. (3,
          4.1.6, 4.3.3).

——— c.   — Sur la déformation infiniment petite d'un ellipsoid élastique soumis à des
          efforts donnés sur la frontière. C.R. Acad. Sci. Paris **133**, 361–364. (3, 4.3.4).

——— d.   — Sur la déformation infiniment petite d'une enveloppe sphérique élastique.
          C.R. Acad. Sci. Paris **133**, 326–329. (4.1.6, 4.3.3, 4.3.4).

1903   Appell, P.: Traité de Mecanique Rationelle, vol. III (esp. pp. 528–532). Paris:
          Gauthiers-Villars. (1, 3).

———   Hadamard, J.: Leçons sur la Propagation des Ondes. Paris: Hermann Cie. (3).

1906   Fredholm, I.: Solution d'un problème fondamental de la théorie de l'élasticité.
          Ark. Mat. Ast. Fysik **2**, Paper No. 28, 1–8. (3, 6.2).

———   Lauricella, G.: Sull'integrazione delle equazioni dell'equilibrio dei corpi elastici
          isotropi. Rend. R. Accad. Lincei (Ser. 5a) **15**, 426–433. (3).

1907   Almansi, E.: Un teorema sulle deformazioni elastiche dei solidi isotropi. Atti
          R. Accad. Lincei Rend. Cl. Sci. Fis. Mat. Natur. (Ser. 5) **16**, 865–867. (3, 7, 7.2).

——— a.   Boggio, T.: Nuova risoluzione di un problema fondamentale della teoria
          dell'elasticità. Atti R. Accad. Lincei Rend. Cl. Sci. Fis. Mat. Natur. (Ser. 5) **16**,
          248–255. (3).

——— b.   — Determinazione della deformazione di un corpo elastico per date tensioni
          superficiali. Atti R. Accad. Lincei Rend. Cl. Sci. Fis. Mat. Natur. (Ser. 5) **16**,
          441–449. (3, 4.1.5, 4.3.3).

———   Korn, A.: Sur un problème fondamental dans la théorie de l'élasticité. C.R.
          Acad. Sci. Paris **145**, 165–169. (3).

1908   — Solution générale du problème d'équilibre dans la théorie de l'élasticité dans
          le cas où les efforts sont données à la surface. Ann. Fac. Sci. Univ. Toulouse
          (Ser. 2) **10**, 165–269. (3, 8.6).

1909   — Über die Cosserat'schen Funktionentripel und ihre Anwendung in der Elastizi-
          tätstheorie. Acta Math. **32**, 81–96. (3).

1925   Weber, C.: Achsensymmetrische Deformation von Umdrehungskörpern. Z.
          Angew. Math. Mech. **5**, 466–468 (5 Appendix).

1927   Love, A. E. H.: A Treatise on the Mathematical Theory of Elasticity, 4th edn.
          Cambridge: Cambridge University Press. Reprinted at New York: Dover
          Publications 1952. (3, 4.1.6).

1928   Trefftz, F.: Mathematische Elastizitätstheorie. Handbuch der Physik, Bd. VI.
          Berlin: Springer. (3).

1930   Weatherburn, C. E.: Differential Geometry in Three Dimensions, vol. II. Cam-
          bridge: Cambridge University Press. (4.4.2).

1933   Muskhelisvili, N.: Recherches sur les problèmes aux limites relatifs à l'équation
          biharmonique et aux équations de l'élasticité à deux dimensions. Math. Ann.
          107, 282–312. (3, 5.1).

1936   Picone, M.: Nuovi indirizzi di ricerva nella teoria e nel calcolo della soluzioni di talune equazioni lineari alle derivate parziala della Fiscia-Matematica. Ann. Scuola Norm. Sup. Pisa (Ser. 11) **5**. (4.2).

1938   Sherman, D.J.: On the distribution of eigen-values of integral equations in the plane theory of elasticity. Acad. Sc. Union S.S.R. Pub. of Seis. Inst. No. 82 [in Russian]. (5.1).

1940   Weber, C.: Zur Umwandlung von rotationssymmetrischen Problemen in zwei-dimensionale und umgekehrt. Z. Angew. Math. Mech. **20**, 117–118. (5 Appendix).

1947   Friedrichs, K.O.: On the boundary value problems of the theory of elasticity and Korn's inequality. Ann. of Math. **48**, 441–471. (8.6).

—   Hove, L. van: Sur l'extension de la condition de Legendre du calcul des variations aux integrales multiples à plusieurs fonctions inconnues. Proc. Kon. Ned. Akad. Wet. A **50** (1), 18–23. (4.1.2).

1950   Fichera, G.: Sull'esistenza e sul calcolo delle soluzioni dei problemi al contorno relativi all'equilibrio di un corpo elastico. Ann. Scuola Norm. Sup. Pisa (Ser. 3) **4**, 35–99. (2.2, 3, 4.2, 8.6).

—   Sneddon, I.N.: Fourier Transforms. New York: McGraw-Hill. (6.3.2).

1952   Green, A.E., Rivlin, R.S., Shield, R.T.: General theory of small elastic deformations superposed on finite elastic deformations. Proc. Roy. Soc. (Ser. A) **211**, 128–154. (2.1).

—   Tiffen, R.: Uniqueness theorems of two-dimensional elasticity theory. Quart. J. Mech. Appl. Math. **5**, 237–252. (5, 6).

1953   Kellogg, O.D.: Foundations of Potential Theory. New York: Dover Publications. (8.1).

—   Muskhelisvili, N.I.: Some Basic Problems of the Mathematical Theory of Elasticity, Third edn. Groningen: P. Noordhoff. (3, 5.1, 5.2, 6, 6.2, 6.3.2, 7.2).

1954   Browder, F.E.: Strongly elliptic systems of differential equations. Ann. Math. Studies **33**, 15–51. (4.1.2, 8.6).

—   Morrey Jr., C.B.: Second order elliptic systems of differential equations. Ann. Math. Studies **33**, 101–159. (4.1.2).

1955   Duffin, R.J.: Continuation of biharmonic functions by reflection. Duke Math. J. **22**, 313–324. (5.2, 6.1, 6.3.1, 6.3.2).

—   Fichera, G.: Sur un principio di dualità per talcune formale di maggiorazione relative alle equazioni differenziali. Atti Accad. Naz. Lincei Rend. Cl. Sci. Fis. Mat. Natur. **19**, 411–418. (8.6).

—   John, F.: A note on "improper" problems in partial differential equations. Comm. Pure Appl. Math. **8**, 494–495. (1).

—   Pucci, C.: Sui problemi di Cauchy non "ben posti". Rend. Acad. Naz. Lincei **18**, 473–477. (1).

—   Sternberg, E., Eubanks, R.A.: On the concept of concentrated loads and an extension of the uniqueness theorem in the linear theory of elasticity. J. Rational Mech. Anal. **4**, 135–167. (1).

1956   Duffin, R.J.: Analytic continuation in elasticity. J. Rational Mech. Anal. **5**, 939–949. (2.3, 4.4.2, 6, 6.1, 6.2).

—   Ericksen, J.L., Toupin, R.A.: Implications of Hadamard's conditions for elastic stability with respect to uniqueness theorems. Canad. J. Math. **8**, 432–436. (1, 3, 4.1.3, 4.1.5).

1957   Ericksen, J.L.: On the Dirichlet problem for linear differential equations. Proc. Amer. Math. Soc. **8**, 521–522. (3, 5.1).

—   Finn, R., Noll, W.: On the uniqueness and non-existence of Stokes flows. Arch. Rational Mech. Anal. **1**, 97–106. (4.3).

—   Hill, R.: On uniqueness and stability in the theory of finite elastic strain. J. Mech. Phys. Solids **5**, 229–241. (1, 8.6).

1958   Diaz, J. B., Payne, L. E.: Mean value theorems in the theory of elasticity. Proc.
            Third U.S. National Congr. of Applied Mechanics, p. 293–303. (7.1).

——   Duffin, R. J., Noll, W.: On exterior boundary value problems in linear elasticity.
            Arch. Rational Mech. Anal. **2**, 191–196. (3, 4.3, 6.1).

——   John, F.: On finite deformations of elastic isotropic material. Inst. Math. Sci.
            New York Univ. Report IMM-NYU 250. (3).

1959   Aleksandrov, A. Ia.: Some relations between the solutions to a two-dimensional
            problem and to an axially symmetrical one in the theory of elasticity, and the
            solution of axially symmetrical problems by means of analytic functions. Dokl.
            Akad. Nauk SSSR **129**, 754–757 [in Russian]. (5 Appendix).

——   Coleman, B. D., Noll, W.: On the thermostatics of continuous media. Arch.
            Rational Mech. Anal. **4**, 97–128. (1, 2.4).

——a.   Signorini, A.: Questioni di elasticità non linearizzata e semilinearizzata. Rend.
            di Mat. (1–2) **18**, 95–139. (1, 7, 7.3).

——b.   — Questioni di elasticità. Statica non lineare. Vincoli unilaterali, statica
            semilinearizzata complementi. Confer. Sem. Mat. Univ. Bari **48**; **49**; **50**. (1, 7, 7.3).

1960   Bramble, J. H.: Continuation of solutions of the equations of elasticity. Proc.
            London Math. Soc. (Ser. 3) **10**, 335–353. (4.1.6, 5.2, 6).

——   Gurtin, M. E., Sternberg, E.: On the first boundary value problem of linear
            elastostatics. Arch. Rational Mech. Anal. **6**, 177–187. (3, 4.3).

——   Movchan, A. A.: Stability of processes relating to two metrics. J. Appl. Math.
            Mech. **24**, 1506–1524. = Prikl. Math. Mekh. **24**, 988–1001. (8.6).

——   Truesdell, C., Toupin, R.: The classical field theories. Handbuch der Physik,
            vol. III/1. Berlin-Heidelberg-New York: Springer. (2.1).

1961   Aleksandrov, A. Ia.: Solution of axisymmetric problems of the theory of elasticity
            with the aid of relations between axisymmetric and plane states of stress. J. Appl.
            Math. Mech. **25**, 1361–1374. = Prikl. Math. Mekh. **25**, 912–920. (5 Appendix).

——   Bramble, J. H.: Continuation of solutions of the equations of elasticity across
            a spherical boundary. J. Math. Anal. Appl. **2**, 72–85. (6).

——   — Payne, L. E.: An analogue of the spherical harmonics for the equations of
            elasticity. J. Math. and Phys. **40**, 163–171. (4.1.5, 4.1.6, 4.3.3).

——   — — On some new continuation formulas and uniqueness theorems in the
            theory of elasticity. J. Math. Anal. Appl. **3**, 1–17. (6, 7.1).

——   — — On the continuation of solutions of the equations of elasticity by reflexion.
            Duke Math. J. **28**, 247–252. (6).

——   Fichera, G.: Il teorema del massimo modulo per l'equazione dell'elastostatica
            tridimensionale. Arch. Rational Mech. Anal. **7**, 373–387. (4.1.7).

——   Finn, R., I-Dee Chang: On the solutions of a class of equations occurring in
            continuum mechanics, with application to the Stokes paradox. Arch. Rational
            Mech. Anal. **7**, 388–401. (4.2).

——   Goldberg, R. R.: Fourier Transforms. Cambridge Tracts in Mathematics and
            Mathematical Physics **52**. (4.1.2).

——a.   Gurtin, M. E., Sternberg, E.: Theorems in linear elastostatics for exterior domains.
            Arch. Rational Mech. Anal. **8**, 99–119. (2.1, 4.2).

——b.   — — A note on uniqueness in classical elastodynamics. Quart. Appl. Math. **19**,
            169–171. (8, 8.1).

——a.   Hill, R.: Uniqueness in general boundary-value problems for elastic or inelastic
            solids. J. Mech. Phys. Solids **9**, 114–130. (3, 4.1.3, 4.1.5, 4.3.2, 5.1).

——b.   — Bifurcation and uniqueness in non-linear mechanics of continua. Problems of
            Continuum Mechanics, Society for Industrial and Applied Mathematics, Phi-
            ladelphia, p. 155–164. (4.1.3).

——   Toupin, R. A., Bernstein, B.: Sound waves in deformed perfectly elastic materials.
            Acousto-elastic effect. J. Acoust. Soc. Amer. **33**, 216–225. (3, 8).

1962 a.   Bramble, J. H., Payne, L. E.: Some uniqueness theorems in the theory of elasticity. Arch. Rational Mech. Anal. **9**, 319–328. (3, 4.3.2, 4.4.2).

—— b.   — — On the uniqueness problem in the second-boundary value problem in elasticity. Proc. Fourth U.S. Natl. Congr. Appl. Mech. 1962, 469–474. (4.1.5, 4.3.2).

——   Hill, R.: Acceleration waves in solids. J. Mech. Phys. Solids. **10**, 1–16. (2.4).

——   Lavrentiev, M. M.: On the Incorrect Problems of Mathematical Physics. Novosibirsk [in Russian]. (1).

——   Mikhlin, S. G.: Multidimensional Singular Integrals and Integral Equations. Moscow: Fizmatgiz 1962. English translation by W. J. A. Whyte. Oxford: Pergamon 1965. (6).

——   Slobokin, A. M.: On the stability of the equilibrium of conservative systems with an infinite number of degrees of freedom. J. Appl. Math. Mech. **26**, 513–517. = Prikl. Mat. Mekh. **26**, 356–358. (8.6).

——   Zorski, H.: On the equations describing small deformations superposed on finite deformations. Proc. of the Internat. Symposium on Second-Order Effects in Elasticity, Plasticity and Fluid Dynamics. Haifa 1962. Oxford: Pergamon 1964. (2.4, 4.1.1, 8.6).

1963 a.   Bramble, J. H., Payne, L. E.: Effect of error in measurement of elastic constants on the solutions of problems in classical elasticity. J. Res. Nat. Bur. Standards (Sect. B) **67**, 157–167. (4.3.2).

—— b.   — — Bounds for solutions of second-order elliptic partial differential equations. Contrib. Differential Eqns. **1**, 95–127. (8.6).

—— c.   — — Some integral inequalities for uniformly elliptic operators. Contrib. Differential Eqns. **1**, 129–135. (8.6).

——   Diaz, J. B., and Payne, L. E.: New mean value theorems in the mathematical theory of elasticity. Contrib. Differential Eqns. **1**, 29–38. (7.1).

——   Ericksen, J. L.: Non-existence theorems in linear elasticity theory. Arch. Rational Mech. Anal. **14**, 180–183. (1, 8.6).

—— a.   Fichera, G.: Sul problema elastostatico di Signorini con ambique condizioni al contorno. Atti Accad. Naz. Lincei Rend. Cl. Sci. Fis. Mat. Natur. **34**, 138–142. (1, 7, 7.3).

—— b.   — Problemi elastostatici con vincoli unilaterali: Il problema di Signorini con ambigue condizioni al contorno. Atti Accad. Naz. Lincei Mem. Cl. Sci. Fiz. Mat. Natur. (Ser. 1 (8)) **7**, (1963/64), 91–140. (1, 7, 7.3).

——   Hayes, M.: Wave propagation and uniqueness in prestressed solids. Proc. Roy. Soc. (Ser. A) **274**, 500–506. (4.1.4).

——   Mills, N.: Uniqueness in classical linear isotropic elasticity theory. M. Sc. Thesis. The University of Newcastle upon Tyne. (3).

——   Movchan, A. A.: On the stability of processes relating to the deformation of solid bodies. Arch. Mech. Stos. **15**, 659–682 [in Russian]. (8.6).

——   Truesdell, C.: The meaning of Betti's reciprocal theorem. J. Res. Nat. Bur. Standards (Sect. B) **67**, 85–86. (2.2).

1964   Bers, L., Schecter, M.: Elliptic equations. Partial Differential Equations. Ed. by L. Bers, F. John and M. Schecter. New York: Wiley. 133–299. (4.1).

——   Ericksen, J. L.: Non-uniqueness and non-existence in linearized elasticity theory. Contrib. Differential Eqns. **3**, 295–300. (1, 8.6).

——   Fichera, G.: Elastostatic problems with unilateral constraints. The Signorini problem with ambiguous boundary conditions. Seminari (1962/63) Anal. Alg. Geom. e Topol. **2**, 1st Nz. Alt. Mat. 613–679. Roma: Ediz. Cremonese. (7.3).

——   Holden, J. T.: Estimation of critical loads in elastic stability theory. Arch. Rational Mech. Anal. **17**, 171–183. (8.6).

1964   Hayes, M.: Uniqueness for the mixed boundary value problem in the theory of small deformations superimposed upon large. Arch. Rational Mech. Anal. **16**, 238–242. (4.4, 4.4.2).

——   John, F.: Hyperbolic and parabolic equations. Partial Differential Equations. Ed. by L. Bers, F. John and M. Schecter, p. 1–123, New York: Wiley. (4.1.1, 8.2).

——   Knops, R.J.: Uniqueness for the whole space in classical elasticity. J. London Math. Soc. **39**, 708–712. (6.2).

1965   Annin, B.D.: Existence and uniqueness of the solution of the elastic-plastic torsion problem for a cylindrical bar of oval cross-section. Pacific J. App. Math. Mech. **29**, 1038–1047. (7.3).

——   Brun, L.: Sur l'unicité en thermoélasticité dynamique et diverses expressions analogues à la formule de Clapeyron. C.R. Acad. Sci. Paris **261**, 2584–2587. (1, 8, 8.3).

——   Ericksen, J.L.: Non-existence theorems in linearized elastostatics. J. Differential Eqns. **1**, 446–451. (1, 8.6).

——   Fichera, G.: Linear elliptic differential systems and eigenvalue problems. Lecture Notes in Mathematics No. 8. Berlin-Heidelberg-New York: Springer. (2.3, 2.4, 8.6).

——   Guha, M.K.: Uniqueness of the problems of equilibrium of an elastic continuum. J. of Technology, Howrah, Bengal Engineering College **10**, 107–114. (4.4.2, 6.6.2).

——   Gurtin, M.E., Toupin, R.A.: A uniqueness theorem for the displacement boundary value problem of linear elastodynamics. Quart. Appl. Math. **23**, 79–81. (8, 8.1).

—— a.   Knops, R.J.: Uniqueness of axisymmetric elastostatic problems for finite regions. Arch. Rational Mech. Anal. **18**, 107–116. (5 Appendix).

—— b.   — Uniqueness of the displacement boundary value problem for the classical elastic half-space. Arch. Rational Mech. Anal. **20**, 373–377. (6, 6.2).

——   Koiter, W.T.: The energy criterion of stability for continuous elastic bodies I, II. Proc. Kon. Ned. Akad. Wet. B **68**, 178–202. (8.6).

——   Shield, R.T.: On the stability of linear continuous systems. Z. Angew. Math. Phys. **16**, 649–686. (2.4, 3, 8.6).

——   Truesdell, C., Noll, W.: The Non-Linear Field Theories of Mechanics. Handbuch der Physik, vol. III/3. Berlin-Heidelberg-New York: Springer. (1, 2.1, 2.4, 3).

——   — Toupin, R.: Static grounds for inequalities in finite strain of elastic materials. Arch. Rational Mech. Anal. **12**, 1–33. Correction, ibid. **19**, 407. (2.4).

1966   Bramble, J.H., Payne, L.E.: Mean value theorems for polyharmonic functions. Amer. Math. Monthly. **73**, 124–127. (6.1).

——   Hayes, M.: On the displacement boundary-value problem in linear elastostatics. Quart. J. Mech. Appl. Math. **19**, 151–155. (2.4, 4.1.2, 4.1.4).

——   Knops, R.J.: Uniqueness of the axisymmetric exterior traction problem in elastostatics. J. Math. Mech. **15**, 187–206. (5 Appendix).

——   — Wilkes, E.W.: On Movchan's theorems for stability of continuous systems. Internat. J. Engrg. Sci. **4**, 303–329. (2.4, 8.6).

——   Mikhlin, S.G.: On Cosserat functions. Problems in mathematical analysis. Boundary value problems and integral equations, p. 59–69. Izdat. Leningrad Univ. [in Russian]. (3).

——   Payne, L.E.: On some non-well posed problems for partial differential equations. Numerical solutions of Non-Linear Differential Equations, p. 239–263. New York: John Wiley & Sons. (1, 8.3).

——   Truesdell, C.: Existence of longitudinal waves. J. Acoust. Soc. Amer. **40**, 729–730. (2.4).

1967   Coleman, B.D., Mizel, V.J.: Existence of entropy as a consequence of asymptotic stability. Arch. Rational Mech. Anal. **25**, 243–270. (1).
——   Gilbert, J.E., Knops, R.J.: Stability of general systems. Arch. Rational Mech. Anal. **25**, 271–284. (3, 8.6).
——   Hill, R.: Eigenmodal deformations in elastic/plastic continua. J. Mech. Phys. Solids **15**, 371–386. (8.1).
——   Lanchon, H., Duvaut, G.: Sur la solution du problème de la torsion élasto-plastique d'une barre cylindrique de section quelconque. C.R. Acad. Sci. Paris (Sér. A) **264**, 520–523. (7.3).
——   Lavrentiev, M.M.: Some Improperly Posed Problems of Mathematical Physics. Springer Tracts in Natural Philosophy, vol. 11. Berlin-Heidelberg-New York: Springer. (1).
——   Lewy, H.: On a variational problem with inequalities on the boundary. J. Math. Mech. **17** (1967/68), 861–884. (7.3).
——   Lions, J.L., Stampacchia, G.: Variational inequalities. Comm. Pure Appl. Math. **20**, 493–519. (7, 7.3).
——   Mikhlin, S.C.: Further investigations of the Cosserat function. Vestnik Leningrad. Univ. (Mat. Mekh. Astr.) **22**, (7), 96–102 [in Russian]. (3).
——   — Maz'ya, V.G.: The Cosserat spectrum of the equations of elasticity theory. Vestnik Leningrad. Univ. (Mat. Mekh. Astr.) **22**, (13), 58–63 [in Russian]. (3).
——   Ting, T.W.: Elastic-plastic torsion problem II. Arch. Rational Mech. Anal. **25**, 342–366. (7.3).
——   Turteltaub, M.J., Sternberg, E.: Elastostatic uniqueness in the half-space. Arch. Rational Mech. Anal. **24**, 233–242. (6, 6.1, 6.2).
1968   Brezis, H.R., Stampacchia, G.: Sur la régularité de la solution d'inéquations elliptiques. Bull. Soc. Math. France **96**, 153–180. (7, 7.3).
——   Coleman, B.D., Mizel, V.J.: On thermodynamic conditions for the stability of evolving systems. Arch. Rational Mech. Anal. **29**, 105–113. (1).
——   Dafermos, C.M.: On the existence and the asymptotic stability of solutions to the equations of linear thermoelasticity. Arch. Rational Mech. Anal. **29**, 241–271. (1).
——   Edelstein, W.S., Fosdick, R.L.: A note on non-uniqueness in linear elasticity theory. Z. Angew. Math. Phys. **19**, 906–912. (4.1.4, 4.1.5, 4.4.2).
——   Green, A.E., Zerna, W.: Theoretical Elasticity, 2nd edn. Oxford: Oxford University Press. (6.3.2).
——   Hayes, M.: On pure extension. Arch. Rational Mech. Anal. **28**, 155–164. (2.4).
——   — Knops, R.J.: On the displacement boundary-value problem of linear elasto-dynamics. Quart. Appl. Math. **26**, 291–293. (8, 8.2).
——   Hill, R.: On constitutive inequalities for simple materials. J. Mech. Phys. Solids **16**, 229–242. (1).
——a.   Knops, R.J., Payne, L.E.: Uniqueness in classical elastodynamics. Arch. Rational Mech. Anal. **27**, 349–355. (8, 8.3).
——b.   — — Stability in linear elasticity. Int. J. Solids Structs. **4**, 1233–1242. (8.6).
——   Turteltaub, M.J., Sternberg, E.: On concentrated loads and Green's functions in elastostatics. Arch. Rational Mech. Anal. **29**, 193–240. (1, 4.2).
——   Wheeler, L.T., Sternberg, E.: Some theorems in classical elastodynamics. Arch. Rational Mech. Anal. **31**, 51–89. (1, 8, 8.1).
1969   Brun, L.: Méthodes énergétiques dans les systèmes évolutifs linéaires. Première Partie: Séparation des énergies. Deuxième Partie: Théorèmes d'unicité. J. de Mech. **8**, 125–166, 167–192. (8, 8.3).
——   Lewy, H., Stampacchia, G.: On the regularity of the solution of a variational inequality. Comm. Pure Appl. Math. **22**, 153–188. (7.3).

1969   Nitsche, J.C.C.: Variational problems with inequalities as boundary conditions
          or how to fashion a cheap hat for Giacometti's brother. Arch. Rational Mech.
          Anal. **35**, 83–113. (7.3).
——   Sewell, M.J.: On dual approximation principles and optimization in continuum
          mechanics. Philos. Trans. Roy. Soc. London (Ser. A) **265**, 319–351. (7.3).
——   Thompson, J.L.: Some existence theorems for the traction boundary value
          problem of linearised elastostatics. Arch. Rational Mech. Anal. **32**, 369–399. (6).
1970   Hill, R.: Constitutive inequalities for isotropic elastic solids under finite strain.
          Proc. Roy. Soc. (Ser. A) **314**, 457–472. (1).
——   Knops, R.J., Payne, L.E.: On uniqueness and continuous dependence in dyna-
          mical problems of linear thermoelasticity. Int. J. Solids Structs. **6**, 1173–1184
          (8, 8.3).
1971   Gurtin, M.E.: Theory of elasticity. Handbuch der Physik, vol. VI, 2nd edn.
          Berlin-Heidelberg-New York: Springer. (8.6).

Subject Index

Springer Tracts in Natural Philosophy